Stiquito™ Controlled!

IEEE
COMPUTER
SOCIETY

IEEE

Stiquito™ Controlled!

Making a Truly Autonomous Robot

James M. Conrad
University of North Carolina at Charlotte

A JOHN WILEY & SONS, INC., PUBLICATION

Published by John Wiley & Sons, Inc., Hoboken, New Jersey.
Published simultaneously in Canada.

For general information on our other products and services please contact our Customer Care Department within the U.S. at 877-762-2974, outside the U.S. at 317-572-3993 or fax 317-572-4002.

Wiley also publishes its books in a variety of electronic formats. Some content that appears in print, however, may not be available in electronic format.

Library of Congress Cataloging-in-Publication Data is available.

ISBN 0-471-48882-8

Printed in the United States of America.

10 9 8 7 6 5 4 3 2 1

Contents

Foreword vii

Preface ix

1 An Introduction to Robotics and Stiquito 1

2 Introduction to Embedded Systems and the Stiquito Controller Board 15

3 PCB Layout and Manufacturing 33

4 Building Stiquito Controlled 63

5 Stiquito Programming Using Texas Instruments MSP430F1122 111

6 A Two-Degree-of-Freedom Stiquito Robot 129

7 Optimizing the Stiquito Robot for Speed 147

8 More Stiquito Controlled 159

Appendix: Sources of Materials for Stiquito 169

Glossary 175

Index 183

About the Authors 187

Foreword

To all my good friends who have an interest in robotics:

In 1992 a small hexapod robot, Stiquito, was announced on the Web. This tiny robot used Flexinol® alloy to actuate its six legs. The result was a walking robot that had an uncanny, silent gait that gave it the appearance of an insect-like form of life. Since then, tens of thousands of these robots have been purchased and many have been built by interested people. Variations on the design that have legs with two degrees of freedom, and that have been powered by "bumper car" power grids, have been constructed by researchers around the world.

Yet most Stiquito robots available to the general public have been educational novelties, lacking the ability to be controlled by a programmable computer. This lack has hampered the use of Stiquito in educational settings, and thwarted the plans of academic researchers who would like to use it as it was originally intended, to study colony robotics and emergent systems. One difficulty has been the low cost of the robot, and the consequent need to design and implement an equally low-cost, low-power, lightweight, and reliable controller.

Dr. James Conrad, my good friend Jim whom I have known for many years now, has taken this project in hand, and succeeded admirably. For the price of a single mobile-wheeled robot of any one of many varieties, you can now build 20 walking Stiquito robots, and use them for months, possibly years with the original Flexinol®. To refurbish them, a yard of Flexinol® will do the trick! The controller and the robot itself are robust if built carefully.

Jim has turned Stiquito into a marvelous research and educational tool, far beyond my original conception of what it could be. I am very proud that he asked me to write a foreword to his book. Stiquito is now Jim's creation, and no better person to bring it to you could be imagined. Jim's expertise in engineering, education, and his passion for research into miniature robotics has taken form in this book. Due to his patient and untiring effort, a new era in robotics has begun.

Welcome to the age of "Stiquito Controlled"!

Warm regards,

Jonathan W. Mills, Ph.D.
Inventor of Stiquito
Associate Professor
Computer Science, Cognitive Science, Program in Neural Science
Indiana University
Bloomington, Indiana 47405

Preface

When Jonathan Mills first envisioned Stiquito, I don't think he had any idea of the impact it would have on schools, universities, and community colleges; or of the excitement it would create in the area of robotics. Since he first published his technical report on Stiquito, we estimate that over 30,000 people have built this robot (at least we know of 30,000 kits that have been distributed and purchased since then).

It was 1992 when Stiquito was first described in a technical report from Indiana University. Since then, two books have been published: *Stiquito—Advanced Experiments* in 1997, and *Stiquito for Beginners: An Introduction to Robotics* in 1999. This third book, *Stiquito Controlled!,* is long overdue. In it are the materials that have been continually requested by Stiquito enthusiasts over the years. In this book, we introduce several changes to Stiquito, especially in its assembly steps. Based on the recommendations of many builders, we have made it easier to build the base robot, while only adding a little bit to the cost.

This book also has a radical feature not seen in many books currently on the market (if any). We have designed and populated a printed circuit board for attaching to the top of a Stiquito robot. This board contains a microcontroller that drives the legs of your Stiquito robot. This circuit board is the result of several iterations of design and testing by many people.

This book can be used on its own for an Introduction to Robotics class or it can be used as a supplement in many different types of classes. Previous Stiquito books have been used in high schools as introduction to technology, in community colleges as an introduction to robotics, and in universities in many different classes: first-year engineering, Introduction to Robotics, Introduction to Bioengineering, Robotics, and Senior Capstone projects. *Stiquito Controlled!* can be used for those courses as well as for an Introduction to Embedded Systems course. How you use this book is, of course, up to you, but there are several suggestions in Chapter 1 for those who may want to use this book in classes.

This book would not have been possible had it not been for the assistance of numerous people. First, I would like to thank Jonathan Mills for the original Stiquito design and continued advice for this book. Jim Martin has again drawn some superb Stiquito cartoons for the beginning of each of the chapters. Several hobbyists, students, and educators wrote parts of some chapters or provided photographs, including: Steven Tucker (Chapter 3), Jonathan Mills (Chapter 4), Rajan Rai (Chapter 5), Andy McClain (Chapter 5), Scott Vu (Chapter 6), Joseph Kubik (Chapter 7), Pam Huntley (Chapter 7), Luke Penrod (Chapter 8),

Serge Caron (Chapter 8), Mark van Dijk (Chapter 8), Christian Lehner (Chapter 8), and Cosmin Rotorio (Chapter 8). Special thanks go to Russ Wanchisen, Su Chol Kim, Sae Kyu Kim, Young Hyoun Sim, and Chae Min Ko for their work on prototypes of the printed circuit board included in this book, and Steven Tucker for the final version. Jeffery Young, Rajan Rai, Andy Mcclain, and Naga Kurnella worked to finalize the software. Robyn Kinney and the Engineers at Texas Instruments reviewed a draft of this book and suggested changes to improve several chapters. Thanks go to the publisher, Angela Burgess, and especially Deborah Plummer for their help in getting this book produced and published (and for their patience!). Many, many thanks go to the reviewers who offered valuable suggestions to make this book better, especially those from my UNC Charlotte Embedded Systems courses, Ken Gracey from Parallax, Inc., and Wayne Brown and Daniel Blair of Dynalloy.

I would like to personally thank my parents, the Conrads, and my in-laws, the Warrens, for their continued assistance and guidance through the years while I worked on the Stiquito books. Also, I would especially like to thank my children, Jay, Mary Beth, and Caroline, and my wife Stephanie for their understanding when I needed to spend more time with Stiquito than I spent with them.

JAMES M. CONRAD

July 2004

Chapter 1

An Introduction to Robotics and Stiquito

Welcome to the wonderful world of robotics and embedded systems! This third book in the Stiquito series will give you a unique opportunity to learn about these fields in a way that has not been offered before. This book may also be the first affordable educational book to describe an autonomous robot and include the robot with the book! This book will provide you with the skills and parts to build a very small robot. It also has a radical fea-

Stiquito Controlled! Making a Truly Autonomous Robot. By James M. Conrad
ISBN 0-471-48882-8 © 2005 IEEE Computer Society

ture not seen in many books currently on the market (if any). We have designed and populated a printed circuit board for attaching to the top of the Stiquito robot. This board contains a microcontroller that drives the legs of your Stiquito robot. This circuit board is the result of several iterations of design and testing by many people.

The star of this book is Stiquito—a small, inexpensive hexapod (six-legged) robot. Universities, high schools, and hobbyists have used Stiquito since 1992. It is unique, not only because it is so inexpensive, but also because its applications are countless. Some examples of uses of Stiquito include the following (with additional sensors and programming):

- Light following or avoidance
- Object detection using infrared or sonar
- Generation of sound or music using a small speaker
- Swarm behavior

This chapter will present an overview of robotics, the origin of Stiquito, and suggestions for how to proceed with reading the book and building the kit.

First, Some Words of Caution

This warning will be given frequently, but it is one that all potential builders must heed. Building the robot in this kit requires certain skills in order to produce a good working robot. These hobby-building skills include:

- Tying thin metal wires into knots
- Cutting and sanding small lengths (4 mm) of aluminum tubing
- Threading the wire through the tubing
- Crimping the aluminum tubing with pliers
- Patiently following instructions

The project requires two to four hours to complete.

Robotics

The field of robotics means different things to different people. Many conjure up images of R2D2- or C-3PO-like devices from the *Star Wars* movies. Still others think of the character Data from the TV show *Star Trek: The Next Generation*. Few think of vehicles or even manufacturing devices, yet robots are predominantly used in these areas. Our definition of a robotic device is: any electromechanical device that is given a set of instructions from humans, and repeatedly carries out those instructions until instructed to stop. Based on this definition, building and programming a toy car to follow a strip of black tape on the floor is an example of a robotic device, but building and driving a radio-controlled toy car is not. Machines requiring continuous human control like battle-bots or R/C cars are not really robots, due to their lack of autonomy.

The term "robot" was created by Karel Čapek, a Czechoslovakian playwright. In his 1921 play, *R. U. R.* (*Rossum's Universal Robots*) [1], humans create mechanical devices to serve as workers. The robots turn on their creators, thus setting up years of human-versus-machine conflicts.

The term "robotics" was first coined by science fiction author Isaac Asimov in his 1942 short story "Runaround" [2]. Asimov can be considered to be the biggest fan of robotics; he wrote more than 400 books in his lifetime, many of them about or including robots. His most famous and most often cited writing is his "Three Laws of Robotics," which he first introduced in "Runaround." These laws describe three fundamental rules that robots must follow in order to operate without harming their human creators. The laws are:

1. A robot may not injure a human being, or, through inaction, allow a human being to come to harm.
2. A robot must obey the orders given to it by human beings, except where such orders would conflict with the First Law.
3. A robot must protect its own existence as long as such protection does not conflict with the First and Second Laws.

These laws provide an excellent framework for all current and future robotic devices.

There are many different types of robots. The classical robots depicted in science fiction books, movies, and television shows are typically walking, talking humanoid devices. However, the most useful and prevalent robot in use in the United States is the industrial arm robot used in manufacturing. These robotic devices precisely carry out repetitive and sometimes dangerous work. Unlike human workers, they do not need coffee breaks, health plans, or vacations (but they do need maintenance and the occasional sick day). You may have seen an example of these robotic arms in auto maker commercials in which an automobile body is welded and painted. Figure 1.1 shows an example of a small robotic arm manufactured by EPSON Robots.

Another type of robot used in industry is the autonomous wheeled vehicle. Wheeled vehicle robots are used for surveillance or to deliver goods, mail, or other supplies. These robots follow a signal embedded in the floor, rely on preprogrammed moves, or guide themselves using cameras and programmed floor plans. They

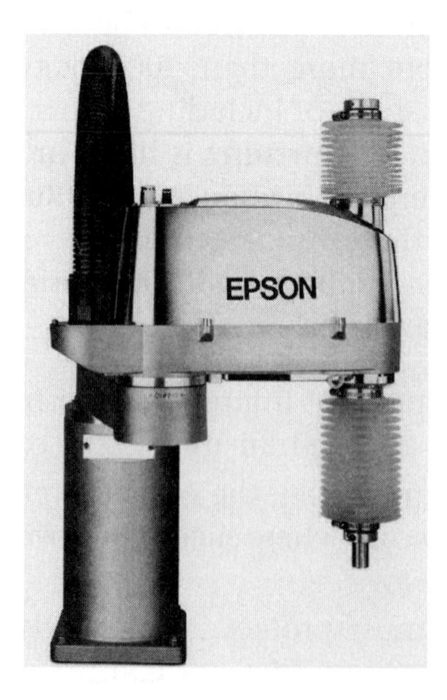

Figure 1.1 Robotic Arm Device. Courtesy of EPSON Robots.

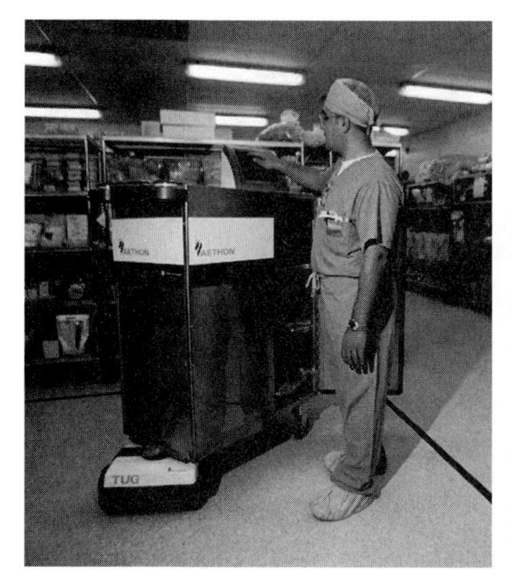

Figure 1.2 The Aethon Tug Autonomous Delivery Robot. (Used by permission of Aethon, Inc.)

usually have object-avoidance hardware and software. An example of an autonomous wheeled robot, shown in Figure 1.2, is the Aethon Tug [3]. This device attaches to a cart and travels through a hospital's hallways, delivering the cart and its contents to a programmed destination.

Although interest in walking robots is increasing, their use in industry is very limited. Still, walking robots have been popular in the "entertainment" market. Walking robots have advantages over wheeled robots when traversing uneven terrain. Two recent entertainment robotic "dogs" are the Tiger/Silverlit i-Cybie (retailing for $200) and the Sony Aibo (retailing for $1600). The i-Cybie (Figure 1.3) was not successful in the market and was discontinued, but has a devoted following. It has a distinct feature of being the lowest-priced programmable entertainment robot. Sony has now produced three generations of the Aibo. They also sell a development kit for reprogramming the robot.

Most walking robots do not take on a true biological means of propulsion, defined as the use of contracting and relaxing muscle fiber bundles. The means of propulsion for most walking robots is either pneumatic or mo-

Figure 1.3 The programmable i-Cybie entertainment robot.

tor-driven. True muscle-like propulsion in inexpensive hobby and educational robots did not exist until recently. A new material, Flexinol®, is used to emulate the operation of a muscle. Flexinol® has the properties of contracting when heated, then returning to its original size when cooled. An opposable force is needed to stretch the Flexinol® back to its original size. This new material has spawned a plethora of new small walking robots that originally could not be built with motors. Although several of these robots were designed in the early 1990s, one of them has gained international prominence because of its low cost. This robot is called Stiquito.

Stiquito

In the early 1990s, Dr. Jonathan Mills was looking for a robotic platform to test his research on analog logic. Most platforms were prohibitively expensive, especially for a young assistant professor with limited research money. Since "necessity is the mother of invention," Dr. Mills set out to design his own inexpensive robot. He chose four basic materials with which to build his designs:

1. For propulsion, he selected nitinol (specifically, Flexinol® from Dynalloy, Inc.). This material would provide a "muscle-like" reaction for his circuitry, and would closely mimic biological actions. More detail on Flexinol® is provided in Chapter 4 and other sources [4].
2. For a counter-force to the Flexinol®, he selected music wire from K & S Engineering. The wire could serve as a force to stretch the Flexinol® back to its original length and provide support for the robot.
3. For the body of the robot, he selected ⅛" square plastic rod from Plastruct, Inc. The plastic is easy to cut, drill, and glue. It also has relatively good heat-resistive properties.
4. For leg support, body support, and attachment of Flexinol® to plastic he chose aluminum tubing from K & S Engineering.

Dr. Mills experimented with various designs, from a tiny four-legged robot, two inches long, to a six-floppy-legged, four-inch long robot. Through this experimentation, he found that the best movement of the robots was realized when the Flexinol® was parallel to the ground and the leg part touching the ground was perpendicular to the ground.

The immediate predecessor to Stiquito was Sticky, a large hexapod robot. Sticky is 9" long by 5" wide by 3" high. It contains Flexinol® wires inside aluminum tubes, which are used primarily for support. Sticky can take 1.5 cm steps, and each leg had two degrees of freedom. Two degrees of freedom means that Flexinol® wire is used to pull the legs back (first degree) as well as raise the legs (second degree).

Sticky was not cost-effective, so Dr. Mills used the concepts of earlier robots with the hexapod design of Sticky to create Stiquito (which means "little Sticky"). Stiquito was originally designed for only one degree of freedom, but has a very low cost. Two years later, Dr. Mills designed a larger version of Stiquito, called Stiquito II, which had two degrees of freedom [5]. A picture of Stiquito II is shown in Figure 1.4.

At about the same time that Dr. Mills was experimenting with these legged robots, Roger Gilbertson of Mondotronics and Mark Tilden of Los Alamos Labs were also ex-

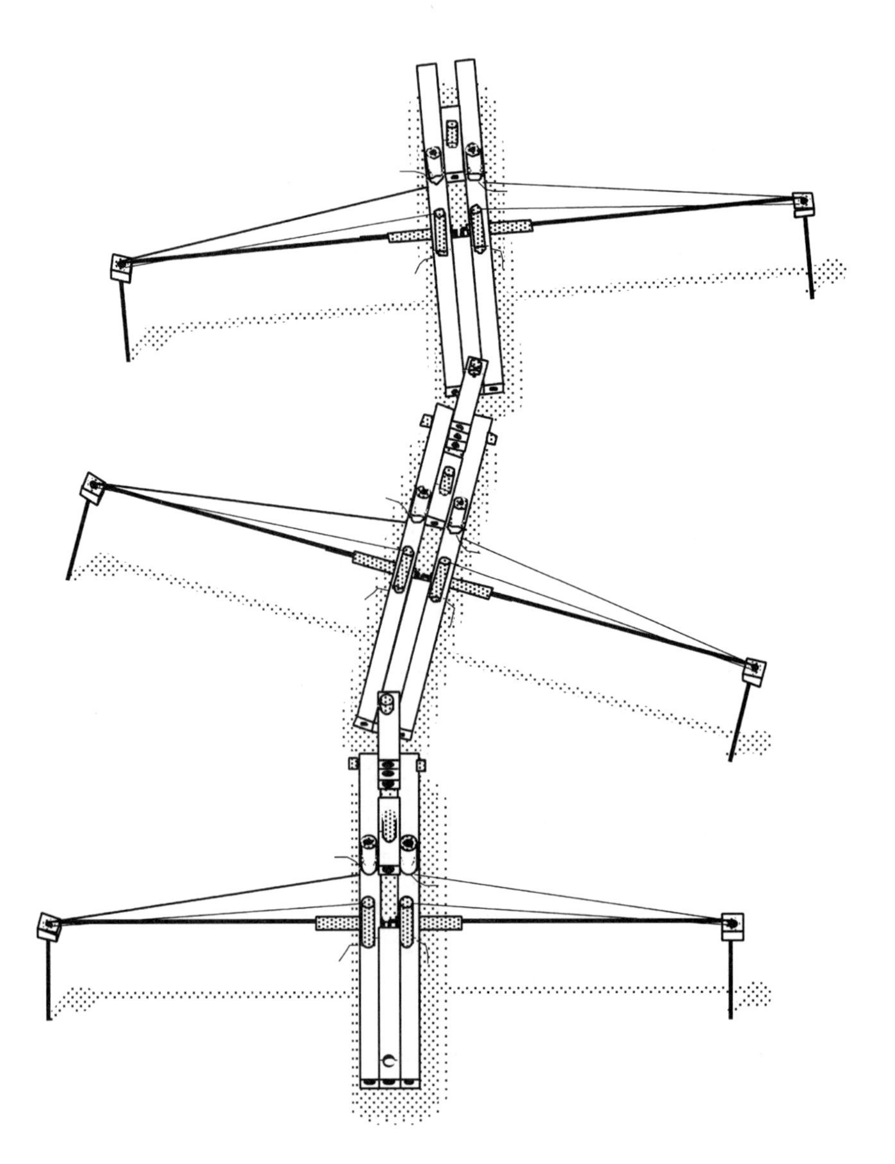

Figure 1.4 The Stiquito II Robot.

perimenting with Flexinol®. Gilbertson and Tilden's robots are described in the first Stiquito book [5].

The original Stiquito kit included in *Stiquito: Advanced Experiments* [5] and *Stiquito for Beginners* [6] relied on aluminum crimps for anchoring the Flexinol® to the legs and body. Through experimentation and user comments, we have changed the assembly procedure to now include screws. This new Stiquito kit is named "Stiquito Controlled."

The Stiquito Controlled Kit

The kit that is included with this book has enough materials to make one Stiquito robot, although there are enough extra components in case you make a few errors while building the robot. The most important thing to remember when building this kit is that Stiquito is a hobby kit; it requires hobby-building skills, like cutting, sanding, soldering, and working with very small parts. For example, in one of the steps, you will need to tie a knot with the Flexinol® wire. Flexinol® is very much like thread, and it is very difficult to tie a knot with it. However, if you have time and patience (and after some practice), you will soon be able to tie knots like a professional.

The kit that is included with this book is a simplification of the original Stiquito described in Jonathan Mills' Technical Report [7] and offered as a kit from Indiana University. In your kit, the plastic Stiquito body has been premolded, so now you no longer have to cut, glue, and drill the plastic rod to make the body. The addition of screws also makes it easier to build and provides the perfect interface to the printed circuit board.

Remember that building Stiquito is not like a Lego® project. This is not a snap-together, easy-to-build kit but a hobby kit, so it takes some model-building skills. Be patient! Make sure all electrical connections are clean and free of corrosion. Sand metal parts before tying, crimping, or attaching them. Allow four hours to build your first robot. Jonathan Mills swears he can build a robot in one hour, but it takes me about two (while watching sports on

TV). This could be a wonderful parent–child project (in fact, my junior-high-school-aged daughter likes to "build bugs with dad"). Make sure to block out enough time to complete the kit.

The intent of this kit is to allow builders to create a platform from which they can start experimenting. The instructions provided in Chapter 2 show how you can create a Stiquito that walks with a tripod gait, that is, it allows three legs to move at one time via control from the printed circuit board. What you should do is to examine your goals for building the Stiquito robot and plans for controlling how Stiquito walks. If your plans include a two-degrees-of-freedom robot, then you should modify the assembly of your robot such that you attach two Flexinol® wires to each leg. If the design of your robot includes adding sensors to the printed circuit board on top, you should consider the function and weight of the added circuitry and programming of the robot.

This book can be used for an Introduction to Robotics class or it can be used as a supplement in many different types of classes. Previous Stiquito books have been used in high schools in an Introduction to Technology class, in community colleges as an Introduction to Robotics class, and in universities in many different classes: First Year Engineering, Introduction to Electrical Engineering, Introduction to Robotics, Introduction to Bioengineering, Robotics, and Senior Capstone projects. This book, *Stiquito Controlled!,* can be used for those courses as well as for an Introduction to Embedded Systems course. How you use this book is, of course, up to you, but there are several suggestions below of which chapters to use for those who may want to use this book in classes. Since this book comes complete with assembly instructions as well as a robot kit, it can easily serve as a required textbook for a class, and only needs a minimal amount of additional electronics necessary to investigate the other areas.

- Chapter 1 provides an introduction to robotics and introduces Stiquito to those who may not be familiar with it. It could be used in all classes since it provides a good background to students.

- Chapter 2 is an introduction to the discipline of embedded systems. It discusses microprocessors, microcontrollers, and complete computer systems. It then discusses the Stiquito controller board specifications and features. It could be used in all classes since it provides a broad assessment of technology central to the control of Stiquito and other robots.
- Chapter 3 is an introduction to the design and manufacture of printed circuit boards. This will be of particular interest to Electrical Engineers and Roboticists, and could be included in courses in these disciplines.
- Chapter 4 shows how to build Stiquito for use with the microcontroller printed circuit board. It also has instructions on how to add other connectors so you can program the board. These new instructions no longer have the manual controller included in the assembly process. Again, it is a central part of the book and could be used in all classes.
- Chapter 5 describes the Stiquito printed circuit board in more detail. It provides hardware-interfacing instructions and programming steps for the Texas Instruments MSP430 microcontroller on the board. This chapter could be used in robotics and embedded-systems courses.
- The Stiquito body is also designed such that all twelve large holes can be used for allowing the legs to have two degrees of freedom (just like Sticky and Stiquito II). Chapter 6 describes another type of microcontroller-based Stiquito that uses two degrees of freedom (lifting legs, as well as moving legs forward) to have Stiquito walk like a real insect. This chapter could be used in robotics and embedded-systems courses.
- Chapter 7 describes how to experiment with different types of leg lengths and assembly designs, as well as gaits. This chapter could be used in robotics courses.
- In Chapter 8 we describe some additional research areas and ideas that you can explore on your own. We provide several short examples of what others

have done with Stiquito. This can be used in all courses.

As with any project conducted in a school setting, you may need additional supplies for those cases in which students break their robot kit. Contact the publisher, John Wiley & Sons, Inc., to purchase additional kits, or contact some of the suppliers listed in the back of the book for repair materials.

Some Final Comments

In your building activities, we cannot stress the importance of following common safety practices:

- Wear goggles when working with the kit. Many parts of the kit act as sharp springs!
- Use care when using a hobby knife. Always cut away from you.
- Use care when using a soldering iron. Watch out for burns!
- This kit is intended for adults and for children over the age of 14.

So what is next, you may ask? Now that you have read the Preface and Chapter 1, you should have a good idea of the content of this book. You need to think of your goals with respect to Stiquito. If you are a student and this book is assigned by your instructor, this process may have been already determined (but you can always do more!). Use the chapters of the book that help you reach your goals. Remember, you can always buy more kits from the publisher if your goal is to build more than one robot.

Bibliography

[1] Čapek, K., *Rossum's Universal Robots,* 1972 (Translated and reprinted).

[2] Asimov, I., *The Complete Robot,* Doubleday & Company: Garden City, NY, 1982.

[3] The home page for Aethon and Tug is http://www.aethon.com.

[4] Dynalloy, "Technical Characteristics of Flexinol®," no date. See also Appendix D of [5].

[5] Conrad, J. M., and J. W. Mills, *Stiquito: Advanced Experiments with a Simple and Inexpensive Robot,* IEEE Computer Society Press: Los Alamitos, CA, 1997.

[6] Conrad, J. M., and J. W. Mills, *Stiquito for Beginners: An Introduction to Robotics,* IEEE Computer Society Press: Los Alamitos, CA, 1999.

[7] J. W. Mills, "Stiquito: A Small, Simple, Inexpensive Hexapod Robot," Technical Report 363a, Computer Science Department, Indiana University, Bloomington, IN, 1992.

Chapter 2

Introduction to Embedded Systems and the Stiquito Controller Board

Introduction to Embedded Systems

Experts say that embedded systems are everywhere around us. Embedded-systems development is the practice of putting small computers in everyday items like microwave ovens, cell phones, and automobiles. One important characteristic of an embedded system is that the device con-

ISBN 0-471-48882-8

tains a microprocessor purchased as part of some other piece of equipment. Other characteristics of embedded systems are that they:

- Typically have dedicated software (may be user-customizable) performing limited, simple functions
- Often replace previously electromechanical components
- Often have no "real" keyboard
- Often have a limited display or no general-purpose display device

The inside of the mobile phone shown in Figure 2.1 is a good example of an embedded system. It contains a microprocessor that executes a limited function (making phone calls), has a simple keypad, and has a small, simple display.

The heart of an embedded system is either a microprocessor or a microcontroller. Both of these electronic devices run software and perform computations. The primary difference is that a microcontroller device typically has a microprocessor *and* other peripheral devices on the same chip. These peripherals can include permanent memory

Figure 2.1 The inside of an Ericsson T60c mobile phone.

(like ROM, EEPROM, or Flash), temporary memory storage (RAM), timing circuitry, analog-to-digital conversion circuitry, and communications circuitry.

Embedded systems is the largest and fastest-growing part of the worldwide microprocessor and microcontroller industry, constituting approximately 99.999% of the worldwide unit volume in microprocessors. This is mainly because the number of microprocessors used in personal computers is small compared to all of the microprocessors and microcontrollers used in embedded systems. Consider that the average home has 30 to 40 processors inside, of which only five are within the home PC. You can find microprocessors and microcontrollers inside televisions, VCRs, DVD players, ovens, and even stove vent hoods [1]. Jim Turley, a writer for *Embedded Systems Programming Magazine,* has predicted that "the amount of processing power on your person will double every 12 months." [2] This bold statement is not so surprising considering that a typical mobile phone now contains three microprocessors, and these microprocessors double their execution speed every year. Analysts also say that embedded systems are in over 90% of worldwide electronic devices, and by the year 2010, there will be 10 times more embedded programmers than other types of programmers.

So, who develops embedded systems? There are many different people who contribute to a final embedded-systems product. These include engineers and scientists who:

- Create the basic technological advances in materials (scientists)
- Design the device enclosure (mechanical engineers)
- Design the circuit boards and components on the circuit boards (electrical engineers)
- Design and write code that interfaces with the user and performs the specific device application (software engineers)
- Design and write software to control the hardware (computer engineers)

- Design the assembly lines that make the devices (manufacturing and industrial engineers)
- Make sure the mechanical, electrical, and computer components of a device work together (systems engineers)

It should be noted that many other people can be considered embedded-systems developers. There are countless students and hobbyist who create embedded systems used in robots, and entrepreneurs who develop the next hot electronic toy.

Product Development

Engineering design is the creative process of identifying needs and then devising a solution to fill those needs. This solution may be product, technique, structure, project, method, or many other things, depending on the problem. The general procedure for completing a good engineering design can be called the "engineering method of creative problem solving."

Problem solving is the process of determining the best possible action to take in a given situation. The types of problems that engineers must solve vary between and among the various branches of engineering. Because of this diversity, there is no universal list of procedures that will fit every problem. Not every engineer uses the same steps in his or her design process. The following list includes most of the steps that engineers use [3]:

1. Identifying the problem
2. Gathering the needed information
3. Searching for creative solutions
4. Overcoming obstacles to creative thinking
5. Moving from ideas to preliminary designs (including modeling and prototyping)
6. Evaluating and selecting a preferred solution
7. Preparing reports, plans, and specifications (project planning)

8. Implementing the design (project implementation)

Often, steps five through eight are expanded into more detailed steps, including:

- Requirements gathering and specification of the entire product
- High-level design (architecture of the system)
- Low-level design (algorithms for the software modules and schematic capture of the electrical circuits)
- Coding and building the hardware (implementation)
- Unit testing of the individual software modules
- Functional testing (building the entire software package)
- Integration testing (putting the software in the hardware)
- Verification and beta testing (making sure the product is compliant with the requirements)
- Ship it!

Engineers developing an embedded system will use these steps to conceive, plan, design, and build their products. The people involved in developing the Stiquito controller board used these steps over several years to create the final product you have in your book.

Microprocessor Basics

This book cannot cover all of the concepts of embedded-system design and microcontroller programming. There are, however, some basic concepts that are important.

The basic concept of an embedded system is electricity. If we ignore the underlying voltage value and just consider the maximum voltage of the system, it is easy to recognize two conditions:

1. Presence of the maximum voltage of the system—we will call this state "1"

2. Absence of a voltage of the system (most often 0 volts)—we will call this state "0"

This basic unit of information is the *binary digit* or *bit*. Values with more than two states require multiple bits. Therefore, a collection of two bits has four possible states: 00, 01, 10, and 11. A collection of eight bits is called a *byte*. Often, we group bits together to represent them in a larger number representation, called hexadecimal. A grouping of four bits is represented by one hexadecimal digit, usually preceded by an "x," or sometimes "0x," as represented in Table 2.1. As an example, binary number 1010 is xA in hexadecimal and 10 in decimal. Binary number 01011100 is hexadecimal x5C.

These values are moved around inside the microcontroller and stored in memory locations called registers. Each register has a unique location that be addressed. These memory locations are in addition to larger stores of useable memory.

Microcontrollers have electrical pins that can be used to send electrical signals out from the device or can be used to accept electrical signals. These electrical pins can be grouped together and called ports. Some port pins are input only, some are output only, and some can be configured to be input or output. If a pin is configurable, then its direction is determined by a "direction" bit. The data of a port can be read from the outside through a port data register. Its direction is set in a port direction register. Often, a port is eight bits wide and is assigned a number, for example, Port 1.

Table 2.1 Hexadecimal representation

Binary	Hexadecimal	Decimal	Binary	Hexadecimal	Decimal
0000	x0	0	1000	x8	8
0001	x1	1	1001	x9	9
0010	x2	2	1010	xA	10
0011	x3	3	1011	xB	11
0100	x4	4	1100	xC	12
0101	x5	5	1101	xD	13
0110	x6	6	1110	xE	14
0111	x7	7	1111	xF	15

Most useful work in an embedded system is based on the input/output of data to/from the microcontroller. This data announces its availability using a mechanism called an *interrupt.* An interrupt is an asynchronous event that suspends normal processing and temporarily diverts the flow of control through an "interrupt handler" routine. Interrupts may be caused by hardware internal to the microcontroller (timer, machine check), by hardware external to the microcontroller (signal from an external communications line), or by software (supervisor, system call, or trap instruction).

The Stiquito Controller Board as an Embedded System

Since 1992, Jonathan Mills and James Conrad have built circuitry that was mounted on top of a Stiquito robot to make it sense its environment and move on its own. These circuits include custom VLSI, popular microcontrollers, and analog circuitry [4, 5]. Hundreds of students, teachers, and hobbyists have since built their own Stiquito circuits as well. Many Stiquito enthusiasts have asked about the availability of a programmable controller. This book and the enclosed controller is the response to their requests.

In order to design a product, one must first identify the requirements of the device. We determined that the Stiquito controller board needed to have the following functionality and fulfill the following requirements:

- The Stiquito controller board shall be designed as inexpensively as possible. All electronic parts on the board are to be the most cost-effective possible, with consideration of materials and assembly. It is anticipated that surface-mount components will be needed.
- At a minimum, the embedded system shall consist of a microcontroller, a transistor driver for the Flexinol® legs, a potentiometer for adjusting gait speed, a connection for power, and two LEDs for output of gaits.

- The printed circuit board shall
  - attach to the Stiquito body
  - be of a common, inexpensive epoxy-resin material with copper plating and solder masking
  - have a prototype area at one end, which will have plated through-holes of 0.035″ diameter, 0.1″ spacing, and be the width of the board, at four rows long
- The microcontroller shall run with a supply voltage of 2.7 to 3.9 volts.
- The microcontroller shall have at least 512 bytes of internal, programmable, nonvolatile memory storage (EPROM, EEPROM, Flash).
- The microcontroller shall have at least 32 bytes of RAM storage.
- The microcontroller shall have the ability to be reprogrammed in the factory and by users, and have the necessary circuitry to support this reprogramming.
- The microcontroller shall have four outputs available for driving Flexinol® legs in a one-degree-of-freedom configuration and a two-degree-of-freedom configuration.
- The microcontroller shall have at least two outputs available for driving LEDs.
- The microcontroller shall have at least one input for determining one- or two-degree-of-freedom operation.
- The microcontroller shall have at least one analog input for the measurement of a potentiometer for determining gait speed.
- The microcontroller shall have at least one A/D converter for this analog input.

The Evolution of the Stiquito Controller Board

The roots of the Stiquito controller board are based on the board developed by Nanjundan Mohon in 1993, and described in *Stiquito: Advanced Experiments* [5]. This board contained a Motorola 68HC11 microcontroller (with EPROM memory), transistors to drive current to each Flexinol® leg independently, and an infrared sensor. The software controlled the legs and accepted signals from the infrared sensor to change the gait. This implementation was only a one-degree-of-freedom controller.

Jonathan Mills observed a two-degrees-of-freedom Stiquito in 1995, and realized that this form of locomotion was needed to make a robot walk quickly. Students had worked on two-degrees-of-freedom robots, but none was based on the Stiquito body until 2001, when two student groups successfully built robots for a class and a race. These robots are described in Chapters 6 and 7 of this book. These robots, though successful, were based on the Parallax Basic Stamp 2 microcontroller, an easy to use but expensive platform. What was needed was a low-cost yet easy to use implementation.

In the Spring semester of 2003, a senior design team at North Carolina State University was given the requirements for a Stiquito controller board, listed above. After investigating low-cost microcontrollers by Microchip (PIC), Renesas, Texas Instruments, and others, they decided that the TI MSP430F1101 was an appropriate device. Through hard work, they created a breadboard circuit for the Stiquito controller board (Figure 2.2), and then created a printed circuit board of the design (Figure 2.3).

Although the design had most of the features listed in the requirements, it did not have the ability to be programmed by users once the microcontroller was soldered to the board. An expert circuit designer took the existing design, added a JTAG interface, made the board smaller, and had a low-cost PCB manufacturer create several boards. This second-generation PCB is shown in Figure 2.4.

Several graduate and undergraduate students tested this second prototype. Through their investigations, they found that the original microcontroller selected (MSP430F1101) had several deficiencies. The first deficiency was that the analog-to-digital (A/D) conversion using the single-slope algorithms was difficult to imple-

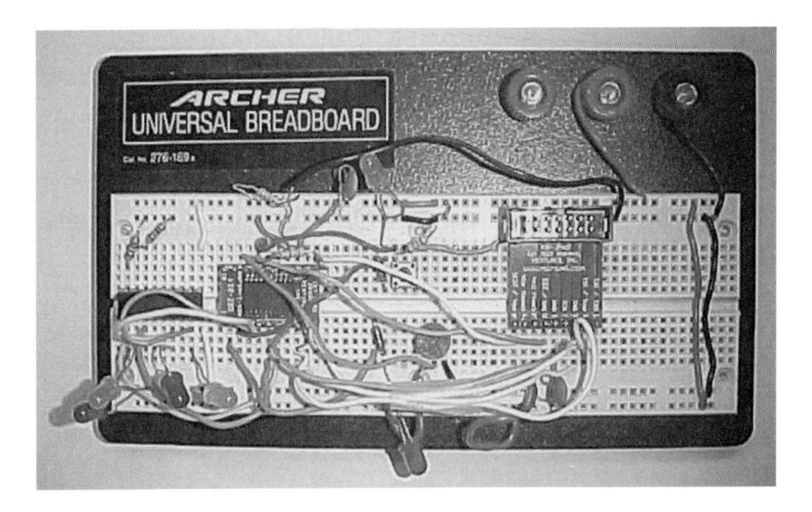

Figure 2.2 A breadboard circuit for the Stiquito controller board.

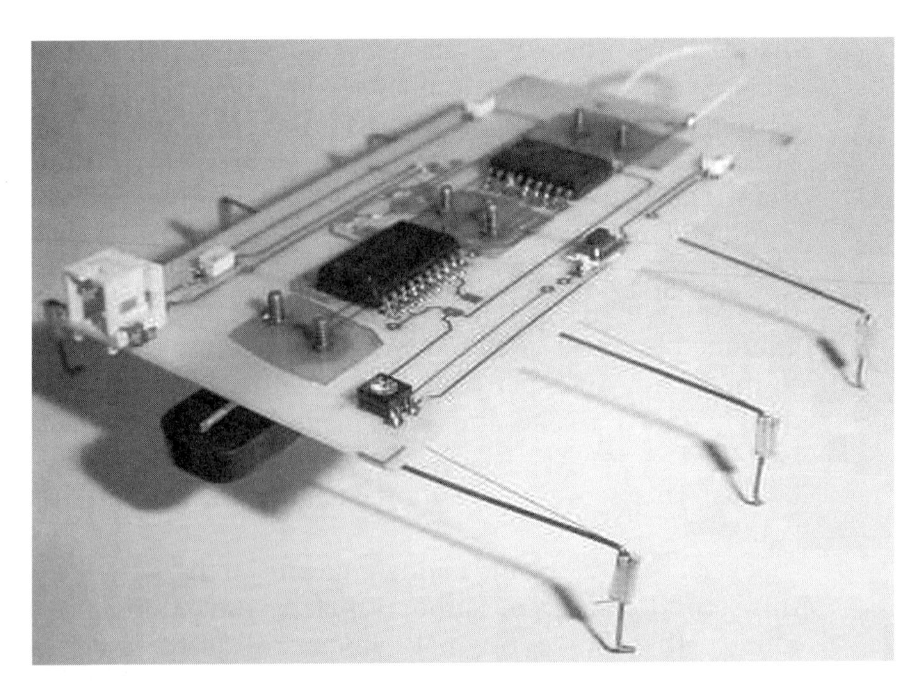

Figure 2.3 The first prototype Stiquito controller board, based on a senior design project.

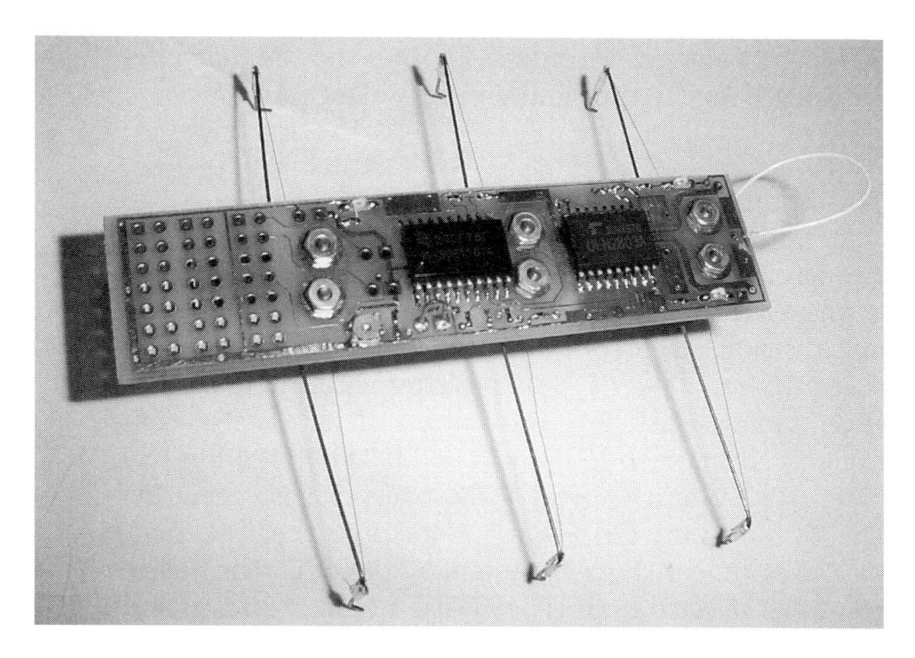

Figure 2.4 A second prototype of the Stiquito controller board made by a low-cost PCB manufacturing company.

ment. Since one of the goals of this board was to allow users to easily develop their own code and flash the microcontroller, we determined that this method of speed control should be changed. A second deficiency was that the microcontroller needed to use three pins and additional hardware to implement the single-slope A/D conversion. Since the number of I/O pins was limited, this would take away some pins that users could use for sensors in the prototype area. The final deficiency was that all of the functionality, including the A/D conversion, one-degree-of-freedom gait, and two-degree-of-freedom gait, barely fit into the 1K byte code space. This would not allow users any extra space to add new software functionality, like the ability to read new sensors they might add to the board.

A new microcontroller, the MSP430F1122, was selected to expand the size of flash memory, to add a true and simple A/D converter, and to add brown-out control. This is a good example of how, during the design process, building a prototype and evaluating the solution showed that a significant change to the requirements was needed.

The Final Stiquito Controller Board

The Texas Instruments (TI) MSP430F1122 microcontroller is one of the parts in the TI family of ultra-low-power, mixed-signal microcontrollers. It has a built-in 16-bit timer; 10-bit, 200-ksps A/D converter with internal reference; sample-and-hold; autoscan; a data transfer controller; and fourteen I/O pins. The MSP430F1122 has 256 bytes of RAM and 4Kbytes + 256 bytes of flash memory. The flash memory can be programmed via the JTAG port, the development kit, or in-system by the CPU. The microcontroller selected for this board has a total of 20 pins and two I/O ports [6]. Port P1 has eight and Port P2 has six I/O signals that are available on external pins. The pin configuration of MSP430F1122 is shown in Figure 2.5.

Apart from the Texas Instruments MSP430F1122 microcontroller, the other main hardware components used for the Stiquito board are listed below.

- **ULN2803AFW transistor (Darlington Sink Driver).** The purpose of this transistor is to amplify the cur-

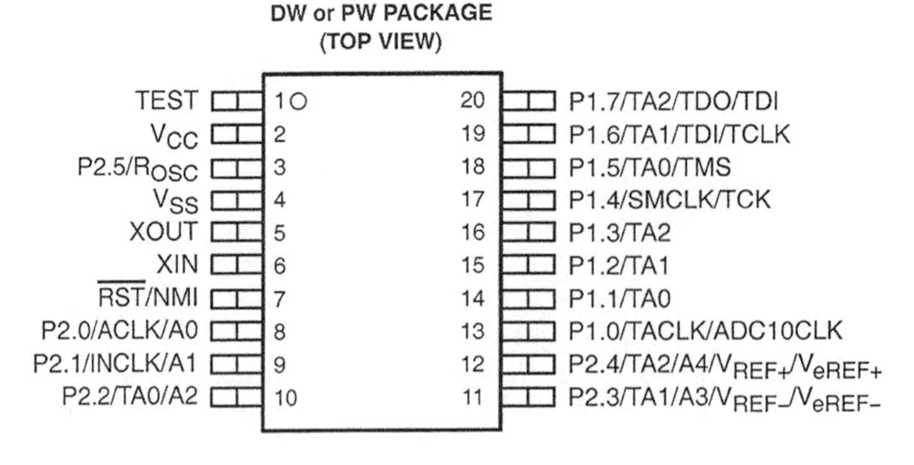

Figure 2.5 The pin assignments of the TI MSP430F1122 microcontroller. (Courtesy of Texas Instruments.)

rent from the microcontroller. The current coming out of the microcontroller is too weak to drive the Flexinol® actuator wire that controls the movement of the Stiquito.

- **LED's.** The purpose of two of the LEDs is to indicate the source output presence and the synchronization of the movement of the Stiquito with the current output. A third LED was added to serve as a power indicator.
- **Potentiometer (POT).** The purpose of the POT is to control the speed of the Stiquito movement. The input voltage is varied by changing the resistance using the potentiometer. The microcontroller then detects the change in voltage to change the speed of the Stiquito.
- **Passives.** Most circuit boards require additional "passive" components, namely resistors and capacitors. Resistors are either used in conjunction with LEDs (to limit the current flowing through the LEDs) or they are used in "pull-up" or "pull-down" configurations. A pull-up or pull-down circuit is used to hold a microcontroller input to Vcc (logic "1") or ground (logic "0"), respectively. The capacitors are used to smooth the voltage changes between Vcc and ground, and are often called "decoupling capacitors."

A schematic of the Stiquito controller board is shown in Figure 2.6. A layout of the components, and their use, is shown in Figure 2.7. Figure 2.8 shows the trace layout of the board, specifically the top layer, bottom layer, and both layers combined.

Although the board that is included in this book does not include the materials to reprogram the microcontroller, several features were added to the board so that users could solder low-cost connectors to the board and be able to expand its capabilities. These features include:

- **JTAG connection.** The JTAG header is connected to a computer using a parallel port interface. A programming adapter is connected between the JTAG con-

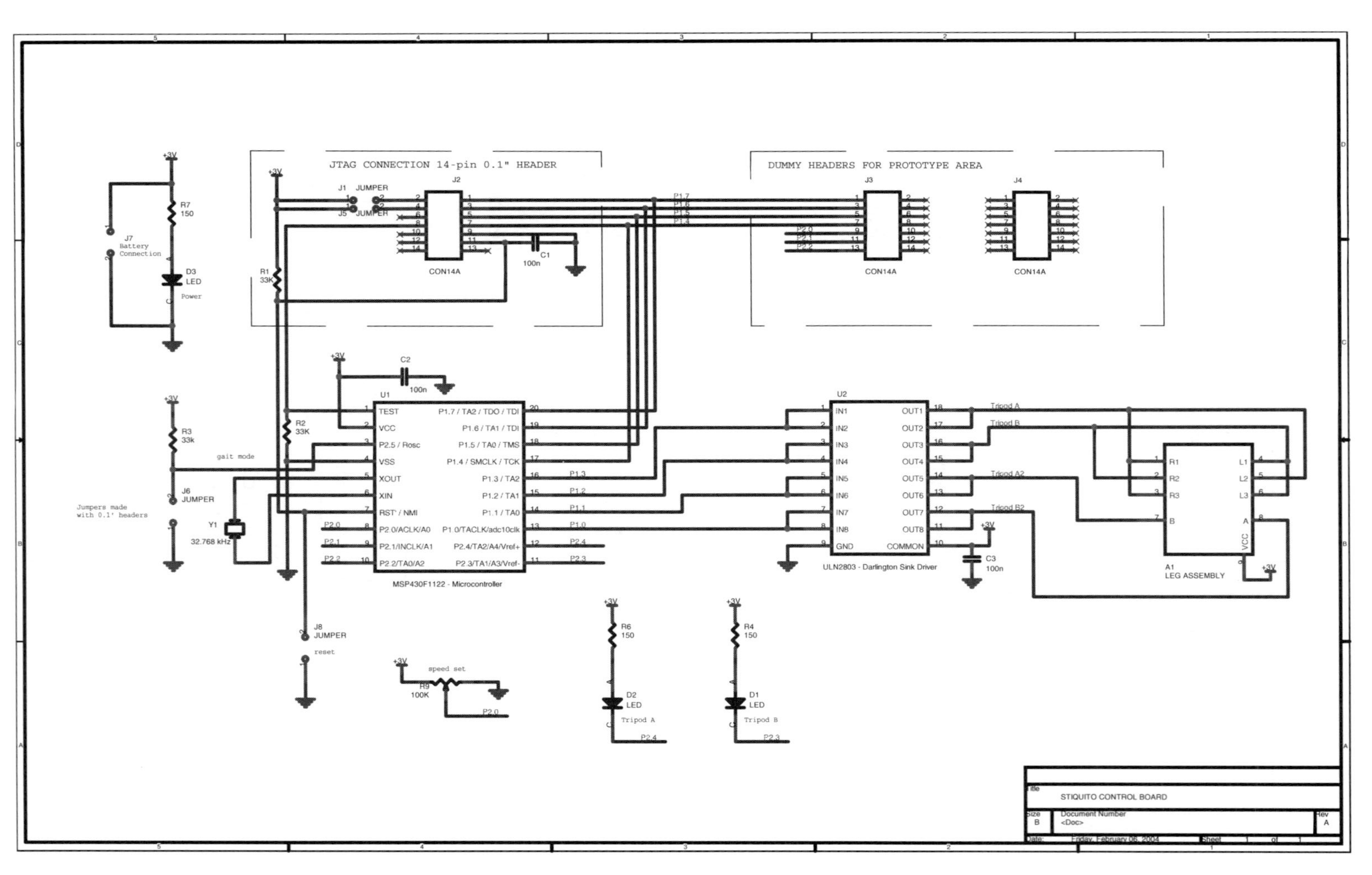

Figure 2.6 The schematic of the Stiquito controller board.

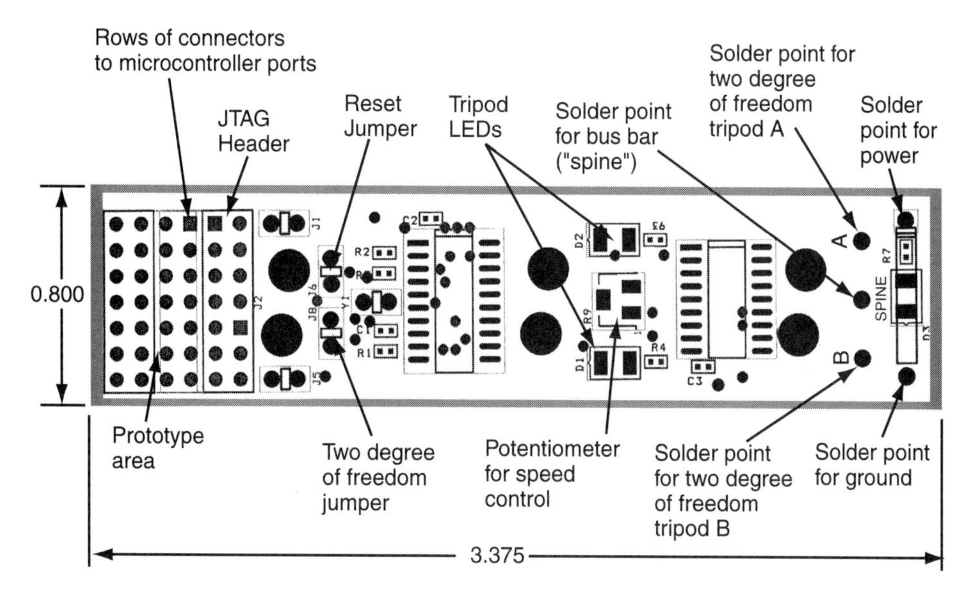

Figure 2.7 The placement and layout of the Stiquito controller board. Board dimensions are in inches.

nector and a PC parallel port. To use the JTAG, you will need to solder a large header and several smaller jumper headers to the board to support powering the Stiquito controller board. This is described in more detail in Chapter 4.

- **Reset Switch.** When a software problem occurs, the reset line can be activated to reset the system to reinitialize the microcontroller. There is a jumper available that can support a switch or jumper header. This is described in more detail in Chapter 4.
- **Watch crystal.** You can add a small low-frequency crystal oscillator to allow for lower power consumption of the MSP430F1122. You will also need to modify the software to support this lower-power mode. This software is not described in this book, but is described in TI datasheets.
- **Prototype area.** A prototype area with through-holes and connections to several microcontroller pins is provided. Specifically, Port pins P2.0 through 2.2,

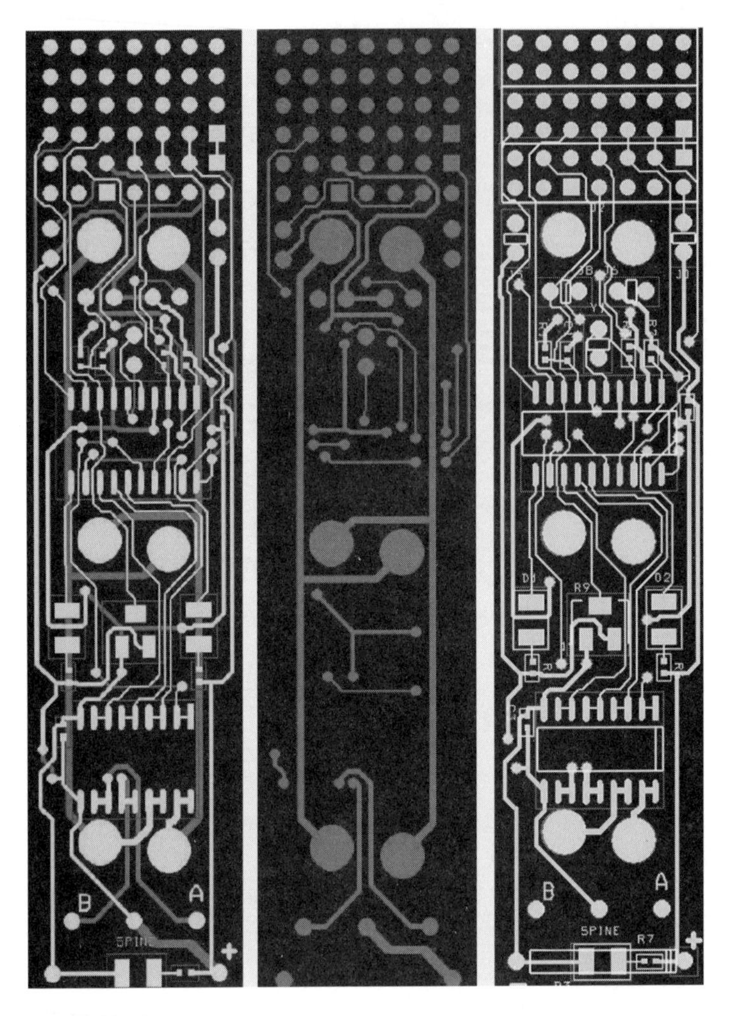

Figure 2.8 The trace routing of the Stiquito controller board.

and P1.4 through 1.7 are connected to one row of through-holes. See Figure 2.7 for more detail.

Chapter 4 will provide additional detail on how to build the Stiquito robot, how to attach the Stiquito controller board, and how to add programming connectors to the board. Chapter 5 will describe how to program the microcontroller.

References

[1] Ganssle, J., "Born to Fail," *Embedded Systems Programming Magazine,* December 2002.

[2] *Embedded Systems Programming Magazine* (http://www.embedded.com).

[3] Wright, P. H., *Introduction to Engineering,* Second Edition, Wiley: New York, 1994, pp. 91–117.

[4] Conrad, J. M., and J. W. Mills, *Stiquito for Beginners: An Introduction to Robotics,* IEEE Computer Society Press: Los Alamitos, CA, 1999.

[5] Conrad, J. M., and J. W. Mills, *Stiquito: Advanced Experiments with a Simple and Inexpensive Robot,* IEEE Computer Society Press: Los Alamitos, CA, 1997.

[6] Texas Instruments, MSP430F1122 Data Sheet, http://focus.ti.com/lit/ds/slas361c/slas361c.pdf.

Chapter 3

PCB Layout and Manufacturing

ISBN 0-471-48882-8 © 2005 IEEE Computer Society

Introduction

Printed circuit boards (PCBs) are the means by which integrated circuit components are connected. The PCB provides structural support and the wiring of the circuit. The individual wires on the PCB are called traces and are made from a flat copper foil. Although many circuit designs may use the same standard components, the PCB is often a custom component for a given design. Not only does the circuit board meet the requirements of electrical functionality, but it is also designed to meet safety requirements.

The simplest PCB has traces on only one side. This is called a single-sided board and is used for very simple and inexpensive circuits. The next level of complexity of printed circuit boards is the double-sided board, which has traces on both sides. Double-sided boards are used for larger common circuits that do not have many digital components. The next-higher level of complexity is the multilayer board, with the four-layer board being the most common. These boards have two outside layers, called the top and bottom signaling layers, which are used for routing and mounting the components. The internal layers are used to create power and ground planes for powering the circuit. The plane layers are solid copper areas with holes in them to allow contacts to pass between the top and bottom layers. Connections between the two signaling layers are called *vias*. If contact is made between the top or bottom signaling layer to one of the plane layers, a special via called a *thermal relief* is used to make the electrical connection. A thermal relief looks like wheel spokes or a star connecting the hole to the plane and is used to make soldering easier. If a thermal relief were not used, the entire plane would need to be heated to melt the solder at a point location. Although this is the normal way a four-layer board is made, there are always exceptions.

Figure 3.1 shows a small section of a PCB that demonstrates some of the new terms introduced above. The figure shows the bottom layer of a double-sided PCB near the location of an 8-pin DIP (dual-in-line package) component. Since this board only has two sides, a ground plane

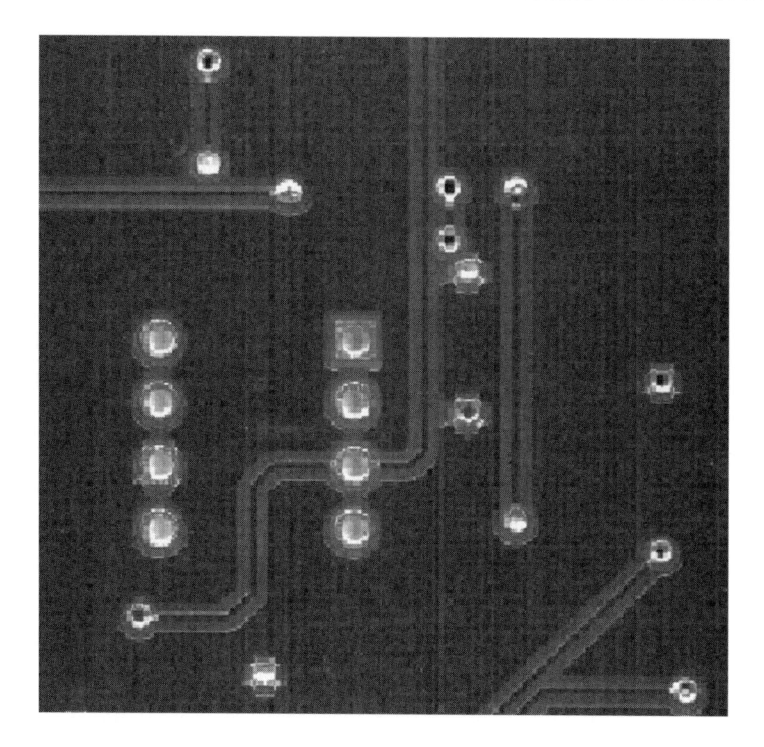

Figure 3.1 A bottom layer of a PCB showing a ground plane, pads, thermal reliefs, and vias.

is used on the bottom layer. Seven silver rings and one square make up the pads of the 8-pin DIP footprint or outline of the component. The pins of the component are soldered to these pads. The square pad is used to denote pin "1" of the component. The dark area is a ground plane. Everything connected to this plane is connected to ground. Looking carefully at one of the pins of the 8-pin DIP, you will notice that it is connected to the ground plane with a cross-like connection of four small traces. This is a thermal relief connecting the plane to the pad. Notice that the ground plane has a fixed spacing from the pads. The thermal relief is made by placing four small traces between the pad and the plane. Notice the trace from one of the pads that is routed to a small hole. This hole is the via that connects the trace to the other side of the board. Now you should be able to spot the other thermal reliefs and vias on this board.

This chapter will describe the design and manufacturing of a PCB and give some details of the Stiquito PCB design. It will address the initial design of the PCB (layout), the physical creation of the PCB (manufacture), and the building of the complete product (population).

Printed Circuit Board Layout

The first step in designing a PCB is to complete the schematic for the circuit design. This schematic is sometimes different from the schematics seen in textbooks. It often contains extra information about the components and construction of the board. For example, a textbook schematic may only show a battery for the power supply, whereas the production schematic would show the battery case, the number of cells, the rating of the battery, and the manufacturers of the components. This information is not needed when designing the functionality of the circuit but becomes important in the assembly of the product. This additional information tells where to order components, documents special features of the components, and helps find problems in reliability and lifetime testing.

In large companies, one team of research designers may develop the initial design to meet the specifications of the new product. The design is then passed to a production team that specializes in completing the design for low-cost production and reliability. Often in small companies, the whole design is completed by the same team or a single individual. Keep in mind that a design is continuously upgraded and changed throughout its lifetime to add features and replace parts as they are updated or discontinued.

At different stages in the design process, the schematic is used to create the PCB. If the board is made during the middle of the design, it is referred to as a prototype board and is used to test and debug areas of the design that are too difficult to test using computer simulation. Prototype boards usually contain special areas to add additional components to correct potential problems with the design.

The PCB made at the end of the design process usually contains options to include extra features that can be chosen when the parts are placed on the PCB. A company may release a line of VCRs, one lower-end model with two video heads and one top-end model with four video heads and hi-fi. Instead of creating two separate PCBs, they would design one board that can function for both. The PCB for the top-end model would have all of the components placed on it, whereas the lower-end model would not have the parts for the extra functions. On the PCB, these would be called open or spare locations.

In the Stiquito design, the main function of actuating the legs has been described in previous books [1, 2]. Extra features were added to the design like speed control, gait-mode selection, a JTAG programming port, and a prototype area. Although these features are mostly programming efforts, they require components on the PCB to provide inputs to the microcontroller. The schematic includes these parts as well as "dummy parts" used as placeholders for the prototype area.

The schematic is made with a CAD schematic-capture tool. This tool allows you to draw a schematic by dragging and dropping part symbols onto a drawing area. The symbols are then connected with a line that represents a wire. Some of these tools are used to simulate the circuit, but for PCB purposes the circuit represents the connections between the components. The final result looks like a standard schematic. CAD schematic-capture tools have a library of part symbols to represent the common components. The library can be searched for the desired part symbol. Smaller libraries can be made to reflect the parts commonly used by a certain company. Manufacturers of components sometimes supply a library of part symbols for their products.

When a part is unique, a symbol must be made for it. The CAD schematic-capture tool has the basic drawing commands for circles, polygons, and text, so making a symbol to represent your unique part is not difficult. When making a symbol in the part editor, you should include a pin for all the connections of the part and clearly label the pins. In the Stiquito design, symbols did not exist for the microcontroller or Darlington transistor driver. This was mostly because there are too many components in the world to have a symbol for each of them. There was also no symbol for the leg assembly because this is a part unique to this product. The symbols for these components were made by drawing a square, adding the correct number of pins, and assigning names to the pins. Figure 3.2 shows the schematic symbol for the Stiquito leg assembly. There is a pin for each leg connection, labeled R1–R3 and L1–L3. There are two pins, A and B, for the second degree of freedom and one more pin for the Vcc spine connection.

In the CAD schematic-capture tool, there are two special symbols that are always used: the power and ground symbols. These symbols represent a global connection to the power supply of the circuit. These symbols may show up many times on the schematic, but the CAD tool realizes that this is a global connection and reports it in the output files. This results in the power supply being connected correctly in the PCB layout.

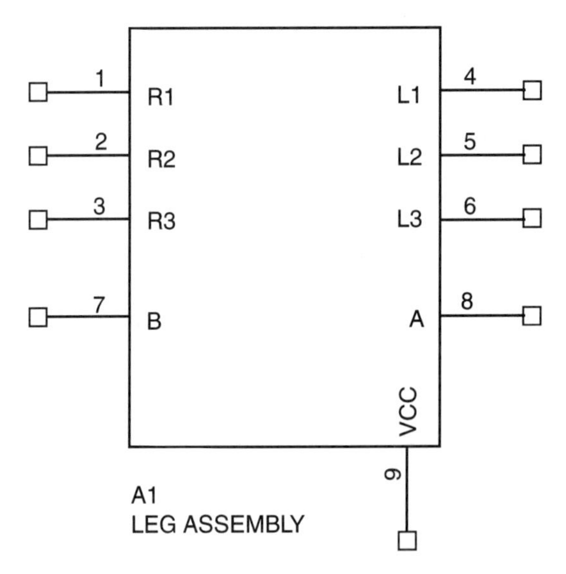

Figure 3.2 Picture of the leg assembly symbol from the CAD schematic-capture tool.

After drawing the schematic, it is checked with an algorithm called a design rule check (DRC). The DRC checks for unconnected pins and components. It also checks for floating wires and other errors specified by the DRC file. The designer has some control over what is checked and can override specific errors if desired. For example, the DRC may warn that a pin is not connected, but the designer can ignore this because the unconnected pin is needed.

Once the DRC is passed, a *netlist* is created from the schematic. The netlist is a text file that describes how the components of the circuit are connected. This file is later used by the CAD layout software. Other reports and files may be generated by the designer as needed. One such file is the bill of materials, commonly known as the "BOM." The BOM contains a list of all the components with values, reference designators, and suppliers of the parts. This information is needed to assemble the components onto your board but is also useful when making the PCB layout. Once all of the reports are generated, it is time to use the CAD layout to design the artwork for your PCB.

Figure 3.3 shows a sample schematic that will be used

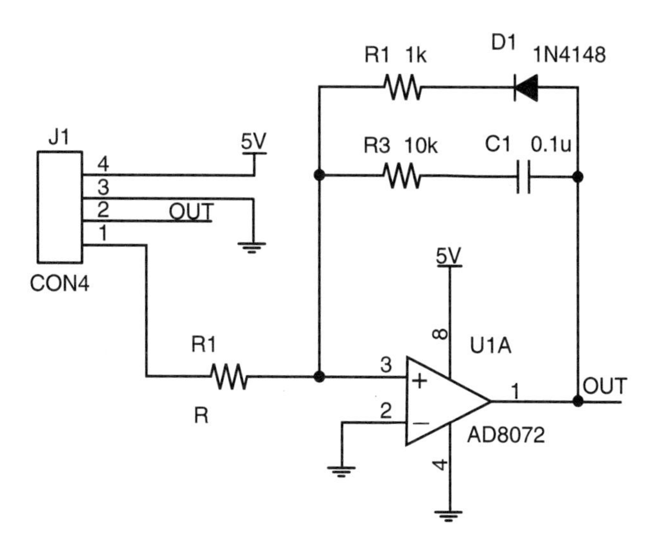

Figure 3.3 Sample schematic from a CAD schematic-capture tool.

to demonstrate the PCB layout. The circuit uses standard components including an op-amp, resistors, diodes, and a connector. At this point, there is nothing designating whether the op-amp is a through-hole DIP package or a surface-mount package. Using the CAD tools database, we can assign a package to the symbol or chose it later. The label "OUT" appears twice on the schematic. The CAD schematic-capture tool treats this connection, or net, as being the same. You will also notice that the resistor R1 has not been assigned a value. This is acceptable for the PCB layout if the footprint is known.

Looking at a PCB on a simple level, it is a set of polygons, circles, holes, and other geometric shapes on different levels of the board. The CAD layout tool allows you to draw these shapes for the different levels in different colors, allowing you to see how they cross over each other and interact. The CAD layout tools also add the benefits of grids, scales, and design-rule checks. Before starting the layout with the CAD layout tool, the designer should collect information about the components used in the design. The information should include data sheets for all of the components, a list of the components specifying which part goes with which reference designator, and the dimensions of everything involved in the design.

When starting a layout project with a CAD layout tool, the project is set up by loading the netlist and footprints of the components. The *footprint* is the physical space that the component uses on the PCB. The footprint contains an outline of the part and a pad for every pin or connection on the component, and tells whether it is a surface-mount or through-hole part. There should be one pad for every pin shown in the symbol used in the schematic. The pin information determines how the pad is made for each layer of the PCB. This is sometimes called a padstack. Padstacks for a surface-mount component would only need a pad on the top or bottom surface. A through-hole padstack would need a pad on the top and bottom surface and also a hole without a pad for the middle layers. The hole in the middle layers, or a plane area, is replaced by a thermal relief if a trace is connected to the plane during routing.

Most CAD layout tools contain a library of footprints for the common surface-mount and through-hole parts like resistors, capacitors, connectors, and integrated components (ICs). Sometimes, a footprint has to be made for special parts with a footprint editor. For the Stiquito design, a new footprint was made for the leg assembly. The Stiquito footprint needed six holes for the screws to connect the leg assembly to the board, two holes for the second-degree-of-freedom connections, and a hole for the Vcc spine connection. The pads needed for the screws should be 160 mils (0.16 inches) in diameter with a 75 mil hole. The six holes are arranged in a 2 × 3 grid. The two rows are 250 mils (¼ inches) apart and the columns are 1031.25 mils ($1\frac{1}{32}$ inches) apart. The rows are offset by 31.25 mils ($\frac{1}{32}$ inches). The unit of mils is used here because it is a standard unit in machining and manufacturing. The measurements for these values come from the mechanical drawing of the plastic leg assembly given in *Stiquito for Beginners* [1]. The other three holes were added to the end of the footprint using drill sizes that were used for the other components to limit the total number of drill sizes used in the project.

A footprint editor is provided in most CAD layout tools to make new footprints for your unique parts. The footprint editor used to create the Stiquito leg assembly footprint had a special grid-array generator. The grid-array generator used for this design was available in the OrCAD Layout Plus™ CAD tool. Figure 3.4 shows the pads being placed by adding dimensions and offsets. Figure 3.5 shows the final footprint generated by the footprint editor and the correct padstacks added to increase the size of the pads.

Sometimes the footprint can be selected for a part in the CAD schematic-capture tool and this information is stored in the netlist. If this is not the case, the footprint must be selected when the netlist is loaded for the new project. It should be remembered that the schematic uses a symbol to represent a part. Different parts can use the same symbol; therefore, different footprints may be assigned to different instances of the same symbol. Two diodes may be shown on the schematic, but one could be

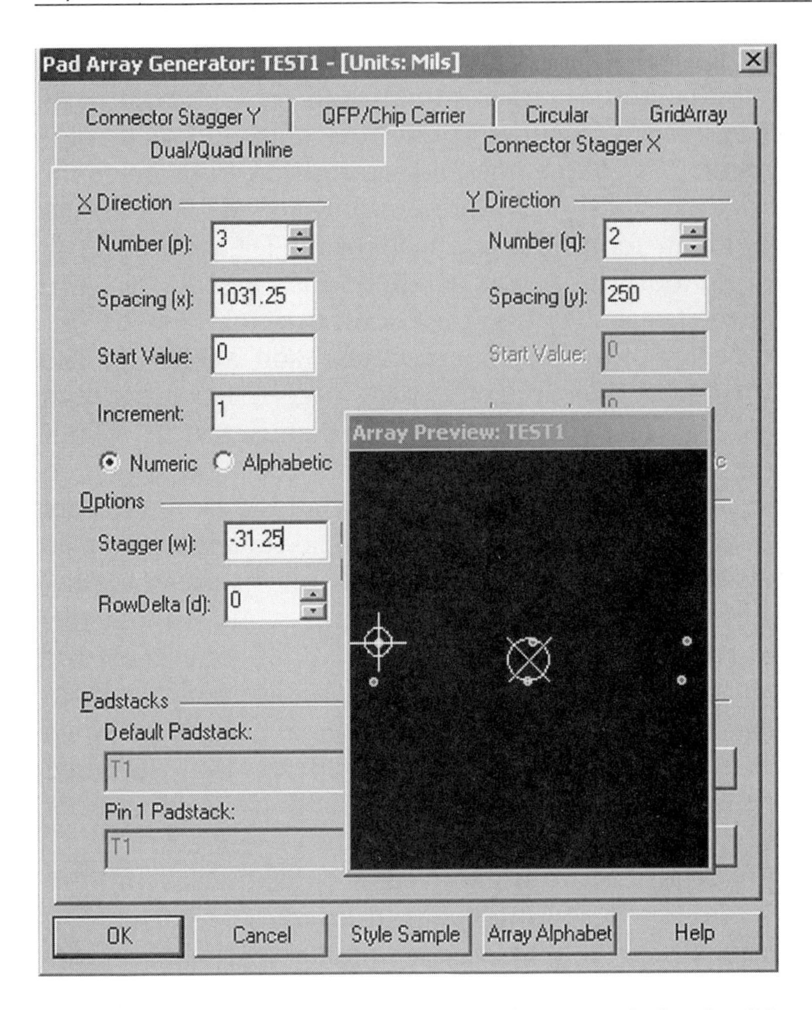

Figure 3.4 The grid-array generator placing pads for the Stiquito leg assembly footprint.

a small surface-mount footprint and the other a high-power component for a power supply.

The placement of footprints begins once the netlist and footprints are loaded. Sometimes, the designer has full control over the shape and size of the board. At other times, the shape and size are determined by mechanical requirements. A circuit board for a PC may have a certain size, shape, and pin combinations to fit within a slot. This information is sometimes in a technology file that contains the board outline and is loaded into your layout pro-

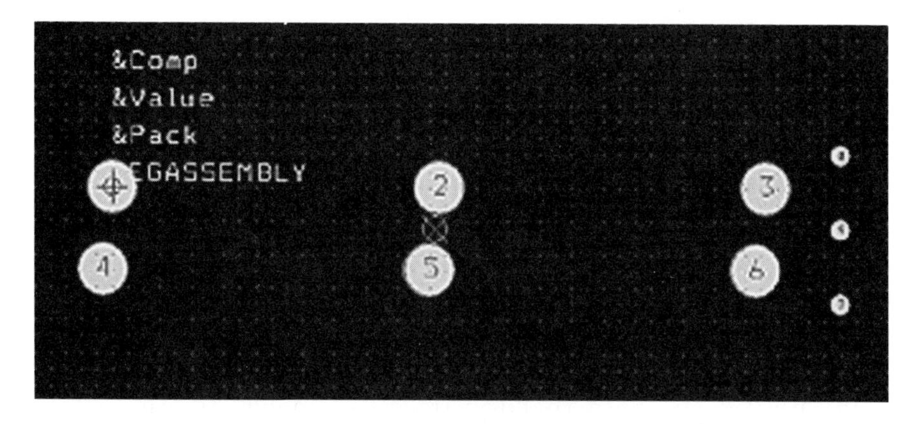

Figure 3.5 The final Stiquito leg assembly footprint.

ject. If there is not a preselected outline for your design, you must make a board outline large enough to hold all of the components and routing. You then shift, rotate, and place the footprints into the board outline. This may seem simple, but careful placement of parts in this stage of design determines how well the traces are routed. Bad placement may make the board unroutable. Here are some general guidelines and tips that are helpful for placing parts:

1. Each component should be placed near other components to which they are connected. This may not be easy for all of the parts.
2. Placing the components in the general layout as on the schematic often helps with the layout process.
3. Decoupling capacitors should be placed as close as possible to the ICs.
4. Components that require heat sinks may require extra room.
5. Components requiring a lot of current may need extra room for routing the power and ground. (This is not a problem in multilayer boards, where contact and thermal reliefs are usually sufficient.)
6. Datasheets for the components may have special placement and routing guidelines to follow for power and noise reasons.

Figure 3.6 shows the component of the sample schematic placed in an outline. The straight lines connecting the pads of one component to another are called the *rat's nest.* The lines of the rat's nest are not a part of the layout but show the work that needs to be done to the board to complete the electrical connections between the pads. The rat's nest also is used to ensure that the correct connections are made in the layout. The designer uses the trace tool of the CAD layout tool to draw the trace. The designer can start at one pad and use the line of the rat's nest to guide him in the path that the trace can make. As one connection is made, the rat's nest line for that connection disappears. When all the lines disappear, all of the connections are made. The rat's nest information comes from the CAD schematic-capture tool and the netlist. Without it, there is a high possibility that a wrong connection would be made. If the schematic is changed during the design, the updated netlist should be loaded into the CAD

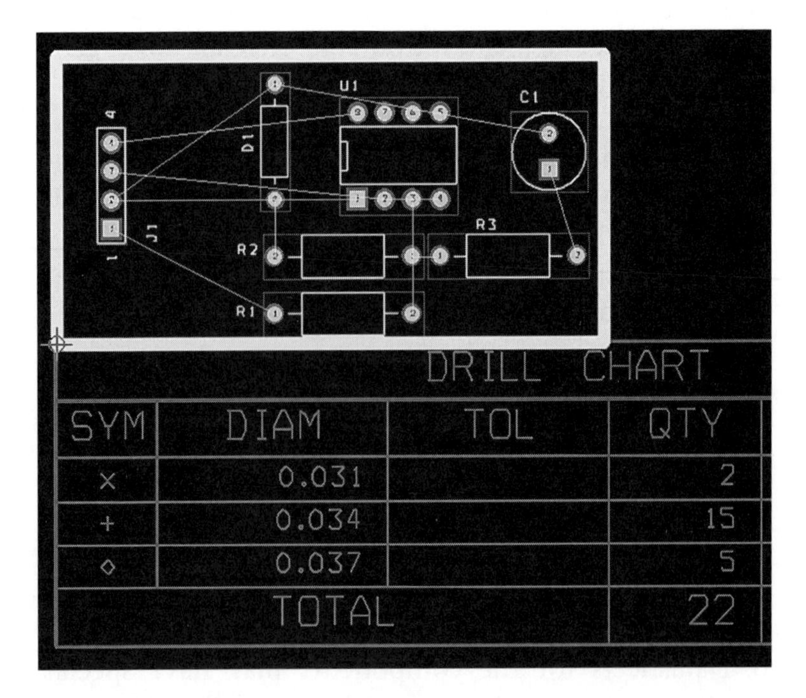

SYM	DIAM	TOL	QTY
×	0.031		2
+	0.034		15
◇	0.037		5
	TOTAL		22

Figure 3.6 The placed footprints of the sample circuit showing the rat's nest.

layout tool to update the rat's nest. The same data that creates the rat's nest also provides some of the DRC rules for the CAD layout tool.

Routing can begin once the components are placed in the board outline. There are two types of trace routing: manual routing done by the designer and autorouting performed by the CAD tool. Manual routing involves drawing the traces between pins that are to be connected, as explained before. Autorouting is done by the computer by running an algorithm to connect the pads of the footprints together. Autorouting usually needs to be "cleaned up" by the designer. The algorithm may not have chosen the most direct path or routed two traces between the same two pins. The designer simply deletes the bad traces and routes them manually. Even though autorouting should be checked, it often saves a great deal of time.

The first step in routing a PCB is connecting the power and ground to each component. Even if you plan to autoroute the design, CAD tools usually recommend that you route the power and ground manually or select them to be routed first by the autoroute algorithm. When dealing with surface-mount components and multilayer boards, the power and ground pins of the components must be connected to the internal layers by a via. Sometimes, you may not be allowed to make a via in the middle of the surface mount pad, depending on limitations of the board house making your PCB. In these cases, the components should be "fanned out" to make the connection. To fan out a component, extend the power or ground pads a short distance with a small trace and connect to power or the ground plane with a via. In some CAD tools, there is an algorithm to do this for you. When routing power and ground connections, you should use thicker traces to carry the required current.

The next step is routing the critical traces. Critical traces may need a certain path, clearance, length, or width to meet design or safety requirements. For these traces, the designer must determine the routing path because the autoroute algorithm is not advanced enough to meet the design requirements. On the Stiquito PCB design, the traces connecting the pads of the leg assembly and Darlington transistor driver were routed before the autorouting because a certain path was desired. After the critical routing, the autorouter is used to complete the final traces. The autoroute traces are cleaned up and a design rule check is completed on the board to check for any missed connections or spacing violations.

Figure 3.7 shows the sample circuit layout with some of the traces routed. Notice that pins 2 and 4 of the DIP socket are connected to the ground by a thermal relief. This is shown by the shape around that pin. Pin 3 of the connector is also connected to the ground plane by a thermal relief. Looking back at Figure 3.3, we see that this is the desired connection. To the right of the four-pin connector J1, there is a trace routed from pin 2 on the bottom layer to a via. The via connects this trace to another trace on the top layer. A short distance later, another via connects the trace to a trace on the bottom layer that is connected to diode D1. Looking at the via, you can see that the two traces overlap. The shaded circle around the via is a clearance region that denotes how close other traces and vias can be placed to the via. It also shows how big to make the holes in the plane layers to allow the via to pass through without shorting out.

When you start the project, you set up rules for the

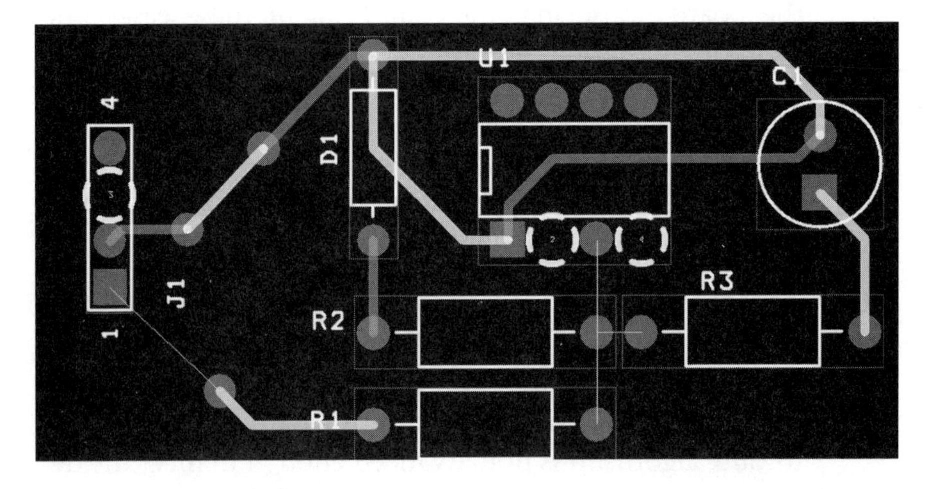

Figure 3.7 Sample layout with some of the traces routed.

spacing of components and traces, as well as the thickness of the traces. The DRC will check your design based on these rules. The rules are made based on the limitations of the manufacturing facility. If they can only make traces with a minimum width of 8 mils, you should make sure that your design rules do not allow you to make 6 mil traces. When the DRC finds a spacing violation, the designer simply moves the footprints and traces around to fix the problem. The designer also fixes trace width violations by making the traces thicker. When working with CAD tools, it is a good rule to run the DRC after manually routing every five traces or so. The DRC will report errors for the traces that are not routed yet, but it catches mistakes early enough to be fixed easily. You should make this a habit, like saving your computer files often. The autoroute routine usually checks the DRC as it is routing, but you should make sure the DRC rules are set to the values that you plan on using.

A silk screen layer is added to the PCB to give the manufacturer, users, and repair technicians part reference designators and other labels that are useful. The silk screen includes text, arrows, and outlines. The silk screen is made on a layer by itself and usually is already included as part of the footprint. After placement and routing, the silk screen text may need to be adjusted and resized to fit close to the parts it represents. Additional silk screen text is added to give copyright, safety, and other information. The silk screen layer can be ignored when manufacturing the PCB by not printing the silk screen on the board.

When the layout is completed and has passed the design rule check, the Gerber files are created for each layer using another algorithm in the CAD layout tool. The Gerber files contain the information for manufacturing your board. There is a file for each layer including, but not limited to, top, bottom, power, ground, top and bottom solder mask, and top and bottom silk screens. Of course, not using some of these in your design lowers the cost. The CAD layout tool also creates a drill file that gives the location and size of the holes.

Figure 3.8 shows the final layout of one of the PCBs de-

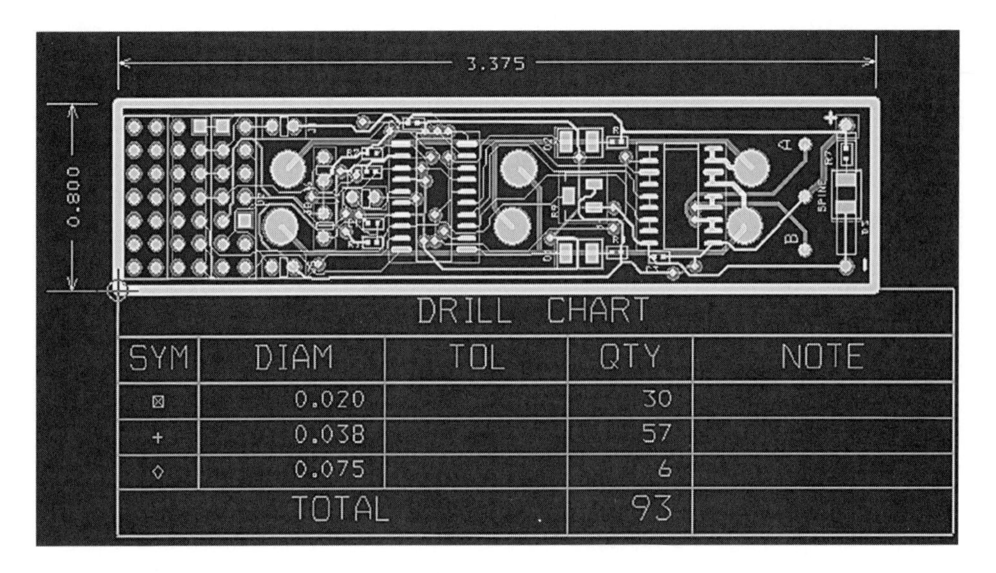

DRILL CHART				
SYM	DIAM	TOL	QTY	NOTE
⊠	0.020		30	
+	0.038		57	
◇	0.075		6	
TOTAL			93	

Figure 3.8 The board layout for a Stiquito PCB.

signed by the authors. The drill chart at the bottom shows the quantity and size of the holes needed for this PCB.

Manufacturing Printed Circuit Boards

In order to understand what is required to have your circuit board made by a PCB manufacturer or board house, you must look at the steps required to make a simple double-sided board. The process explained here is only an introduction. Advances in technology have led to the development of different tools and techniques to perform the same job. A board house may use different techniques for boards of different complexity and quantity and also reorder the processes to produce different results. The processes described here are also presented in general terms; each process is composed of a number of smaller steps including cleaning, inspecting, and curing the different finishes on the board.

When you provide the PCB artwork (Gerber files) to the board house, they check the files for errors and use them to create a series of masks and stencils. The masks are made of a film transparency that is used to expose some areas to light and block exposure in other areas. These masks have only clear and black areas, no gray tones. The stencils are made from a metal foil with patterns cut into it to allow paint or other fluids to pass through in certain areas. PCB manufacturing is a chemical process in which certain areas are affected by one chemical and not by others. The correct combination of chemicals produces the desired results.

The PCB begins with the raw board material called a *laminate*. The laminate is made of structural material laminated with copper on both sides. The structural material can vary depending on the application. A common laminate is called FR-4, a fiberglass–resin material that is lightweight and inexpensive. Other materials are used for projects that require special specification. For example, some laminates have specific dielectric constants for high-frequency circuits. The laminate is cut into a panel of a certain size based on the manufacturer's machinery. Most of the time, the same design is printed multiple times to use the entire sheet and the PCBs are cut apart as the final step. Figure 3.9 shows a panel containing four PCBs for a mobile phone.

For a standard double-sided PCB, the first manufacturing step is drilling the holes for the vias and the through-hole components. Through-hole components have their leads inserted into the PCB through a plated hole. A point in the panel is labeled as the zero reference, and a computer-controlled machine moves the drill to various hole locations based on the data provided in the artwork. Several panels are stacked together and drilled at once. Many drill sizes are available, but limiting the number of drill sizes used on the design can reduce the cost. The panels are cleaned and deburred to remove any material remaining.

The drilled panels are coated with chemicals to enhance the electroless plating. Electroless plating is a chemical process that does not use electricity to adhere the metal to a surface. This type of plating is needed since the inside of the drilled hole is the structural material of the laminate and is not metal. In this process, the panels are placed into a solution that plates the inside of

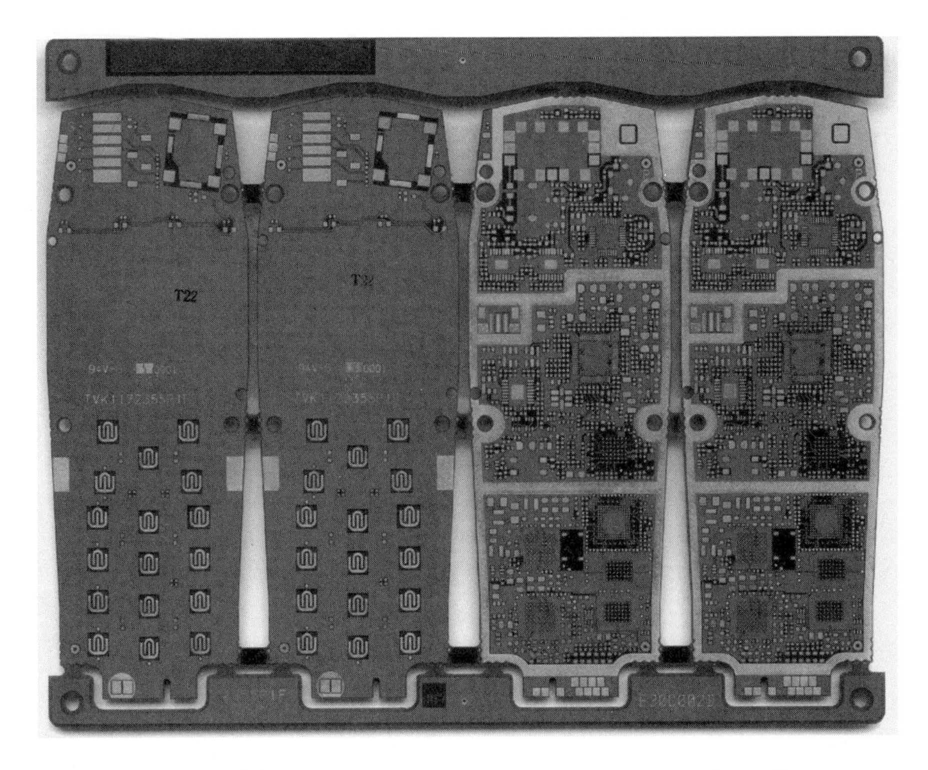

Figure 3.9 A panel containing four PCBs for a mobile phone.

the drilled holes with copper. A ring of copper is formed in the hole, making the electrical connection between the two sides. Any holes that are not plated are plugged for this process or can be drilled at a later time. This process is followed by electroplating which increases the thickness of the copper laminate. The plating of the original laminate is usually thin and may not handle the current and voltage requirements of the design. Nickel, silver, and gold plating are also available from some PCB manufacturers.

Once the panels are drilled with holes and the holes are plated, the trace pattern can be made. This process is similar to photography. The artwork files are used by a photoplotter to make the artwork pattern on a transparency or mask. This mask is used to put the circuit pattern on the panel. The panel is coated with an ultraviolet-light-sensitive chemical called a *photoresist*. The mask is placed over the panel, and the assembly is exposed to a high-wattage light source. The mask is removed, and the panel is put through a series of processing chemicals. The developing chemicals change the properties of the photoresist based on its exposure to the ultraviolet light. The exposed areas of photoresist harden and become resistant to the etching process. The panel is placed in an acid bath and the unexposed areas of copper are etched away. Care must be taken when etching boards with extremely small traces or a mixture of large and small traces.

After the etching process, the panels are cleaned, and the result is a board with traces on both sides and holes that are plated to connect the two sides. The boards could be used for a circuit at this point if you manually placed the components, but most of the time, processes are added to make the PCBs usable for automated assembly and increased reliability. To do this, the next step is to coat the panel with a *solder mask*. The solder mask is like a thin lacquer that covers the board in areas that need to be protected from solder. Without the solder mask, solder would stick to all exposed copper during the component assembly. With the solder mask in place, the solder will only stick to the exposed areas needed to connect the components to the PCB. There are different wet and dry processes used to apply the solder mask. One method is to coat the panel with the liquid lacquer and dry it in an oven. A photoresist and film are again used to define areas to be removed. The artwork for the solder mask pattern is usually defined in the padstack and created automatically with the other Gerber files. The solder mask is then removed from the areas in which it is not desired. The solder mask is finally cured in an oven to fully harden it. The solder mask is usually green and gives the PCBs their distinctive color, but some manufacturers now offer colors like red, blue, yellow, white, and black.

The silk screen, or legend, is now applied to the board using a screen stencil. The silk screen is usually printed in white or yellow ink. The PCB is baked again to harden the silk screen. Figure 3.10 shows part of a PCB that has silk screening ink applied. The silk screen gives the rating of

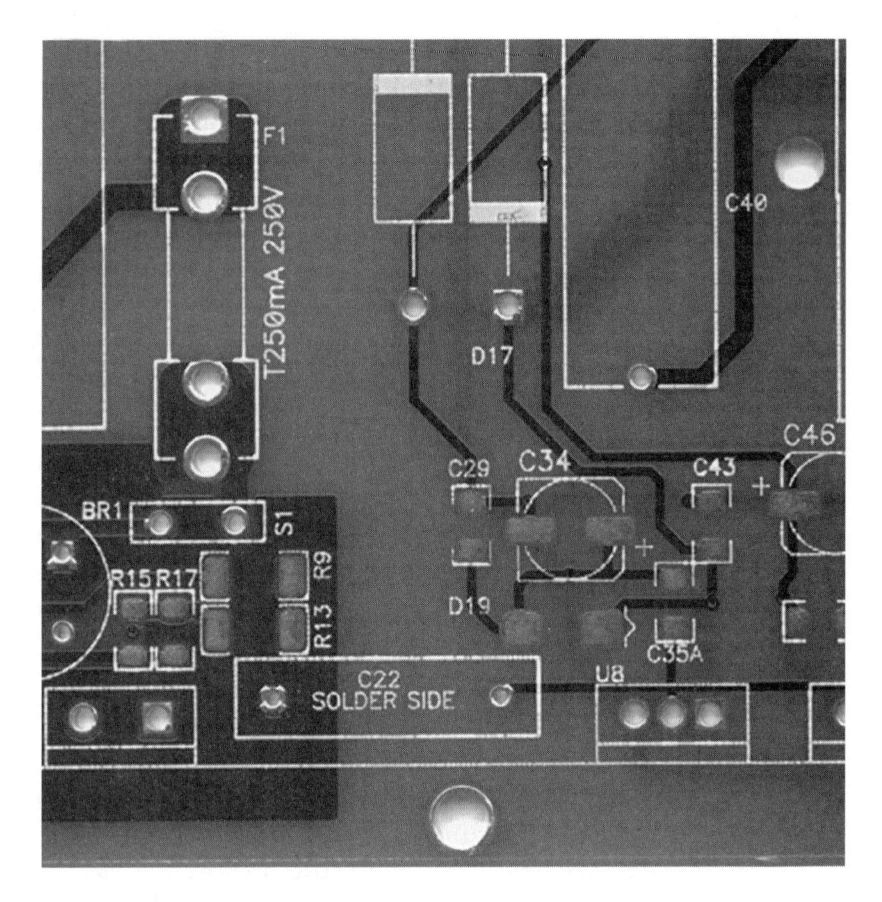

Figure 3.10 A section of a PCB showing the silk screen details.

the fuse F1, reminds the manufacturer that capacitor C22 is mounted on the solder side, denotes which end of diode D19 is the cathode, and denotes which end of the electrolytic capacitor C34 is the positive end.

At this point, the panels have a solder mask, a silk screen, and exposed copper pads. The pads need to be tinned to allow good solder flow during the assembly process. The pads are coated with tin, solder, silver, or even gold. This process can be completed by immersing the board in liquid solder. The more advanced machines for this step spray the panel with *flux,* a chemical that helps the adhesion of solder, then dips the board into a solder tank. The machine then blasts the panel with hot air to move the solder into the holes and smooth the surface of the solder.

The final step is cutting the panel into the individual circuit boards. This is sometimes skipped for smaller boards until the components are added, in order for the board to fit in the assembly machinery (seen in Figure 3.9). Production boards are often electrically tested after they are made to check for shorts or opens that would be difficult to find after the components are added to the board.

A multilayer board is made simply by joining boards together before drilling. The inner layers are etched onto the panels similar to the method used in etching the traces on the top and bottom layers. The panels are cleaned and coated with an adhesive. The different layers are stacked with an insulating layer between them. The stack is then placed in a hydraulic press to adhere the separate panels together. The result looks like the double-sided panels ready for drilling. Figure 3.11 shows the formation of a multilayer board.

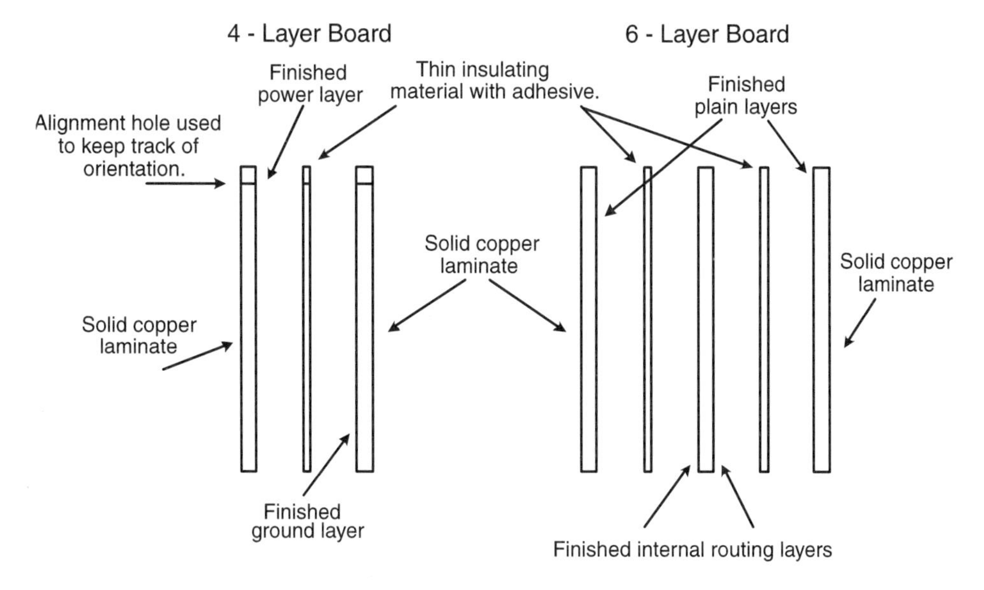

Figure 3.11 Diagram explaining the assembly of a multilayer board.

Populating a Printed Circuit Board

The PCB assembly process uses information developed in the layout process. An important stencil is made from the artwork and used as a mask for the solder paste. Other information about the components is stored during the design process, including the values of the components for a given reference designator.

Placing components on the PCB is called populating the board and is usually performed by a contract manufacturer. The first step is placing the surface-mount components on the PCB using an automated process. The pads of the finished PCB are coated with solder paste using the solder paste stencil. Solder paste has the consistency of toothpaste and holds the component in place when it is placed onto its pad. The stencil covers the entire PCB but has small holes over the surface-mount pads. The solder paste is squeegeed onto the board and covers only the exposed surface-mount pads. The parts are then placed on the board by a robot. The robot picks up the components from trays (for the larger parts), or from tubes (for the smaller parts), or from reels (for large and small parts). The orientation and location of the components are determined by the data provided by the layout files. The boards now have the surface-mount components held in place by the solder paste, but they can still be easily moved out of place. The boards are stored horizontally in trays to avoid such disturbances. In some cases, glue is also used to hold the parts, especially when the PCB has surface-mount components on both sides.

The boards are then loaded with the through-hole components. The axial leaded component, like resistors, diodes, and some other components, can be placed by a machine. Other components are placed by workers in an assembly line. The leads of the components are trimmed to a designated length. The boards are returned to the tray for storage until the next process.

The fragility of the loose components is soon corrected by the next step. The boards are placed on a conveyer belt and passed through a reflow oven and solder wave.

Figure 3.12 A populated panel of twenty Stiquito controller boards.

The oven has several chambers through which the board travels for a few minutes. The first chamber sprays the board with flux. The board is then slowly heated to prevent warping. At some point, the board becomes hot enough to melt the solder paste and solders the surface-mount components to their pads. The board then passes over the *solder wave*. The solder wave is a tank of melted solder that has a small ripple flowing over the surface. As the board passes over the solder wave, the ripple touches the bottom of the board. The properties of solder cause it to be wicked up into the holes. This is the stage in which the solder mask is critical. After the solder wave, the board is cooled at a slow rate, again to reduce warping.

The next step is to wash the board to remove the flux. Flux is needed in the soldering process but can cause long-term problems with the board if not removed. Some parts are designated "no wash." These parts are added to the board with special solder after the washing process to prevent getting them wet.

Figure 3.12 shows a populated panel of twenty Stiquito controller boards just before the individual boards are broken from the panel.

Testing a Printed Circuit Board

The completed boards are electrically tested for functionality. Any board failing the test is inspected in more detail to find the problems. These boards are repaired to increase the passing rate acceptable to the manufacturer. The boards are wrapped in antistatic materials and shipped to the next location for the next stage of production.

We can examine the Stiquito controller board as an example of testing a populated PCB. In order to functionally test the board, one needs to create a device that will temporarily attach to the board and make electrical connections. This device is called a *test fixture*. Often, the test personnel will place each board into a waffle-iron-like device, close the lid, and allow the test circuitry to exercise all the different parts of the circuitry of the board under test.

In a test fixture, several large alignment pins are placed in exact locations in a plastic board. These large pins will serve as guides to place the board under test. Several smaller pins with spring-loaded tips are also placed into the board. These pins make electrical connections to the board under test when it is placed in the fixture. Then a lid is lowered to hold the board under test in place, and the test is started.

The test fixture for the Stiquito controller board is rather simple. It consists of four alignment pins, twenty-five test pins, four LEDs, four resistors, two switches (one momentary, one toggle/slide), one TI programmer, 3 V power supply, and a base test enclosure. Since there are seven LEDs that must be observed and one potentiometer that must be turned, the cover on the fixture is made so as to not cover the board under test. In order to test the Stiquito controller board, the steps in Table 3.1 should be followed. Figure 3.13 shows the circuitry that was wired for

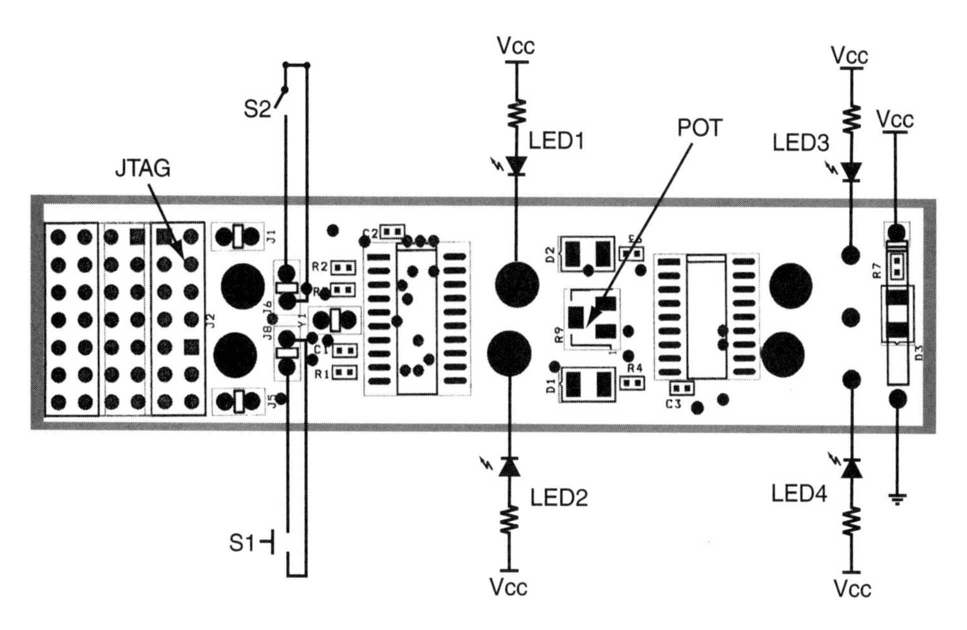

Figure 3.13 A circuit diagram of the test fixture for the Stiquito controller board.

Table 3.1 Stiquito controller board manufacturing test steps

Manual operation	Time (sec)	Additional function verified with this step	Additional component verified with this step
Insert board into test fixture (close lid).	5		
Download the program to the MSP430 flash, read back. Observe Power LED.	10	Board can be powered up (no major shorts of power and ground)	MSP430 (flash, JTAG interface, pins 1, 2, 4, 17–20); D3, R2 R7, C1, C2
Press Reset button (S1: momentary switch).	1	MSP430 Reset	MSP430 (reset, pin 7), R1
Visually check test fixture LED 1 and 2 and Stiquito controller board LEDs.	10	Software operates, ULN drives LEDs	MSP430 pins 11, 12, 15, 16; D1, D2, R3, R4, R6, C3; ULN pins 1–4, 9, 10, 15–18
Turn POT counter-clockwise to observe slower flash; turn POT clockwise to observe faster flash. Return POT to middle position.	20	Software operates, POT turns and works	MSP430 pin 8, R9
Press 2d of button (S2: toggle or slide switch).	1		
Visually check test fixture LED3 and 4.	10	Test ULN drives LEDs	MSP pins 13, 14; ULN pins 5–8, 11–14, R3
Remove Stiquito controller board from test fixture.	5		
Total time	62		

the factory test fixture. A photograph of a prototype of the text fixture is shown in Figure 3.14.

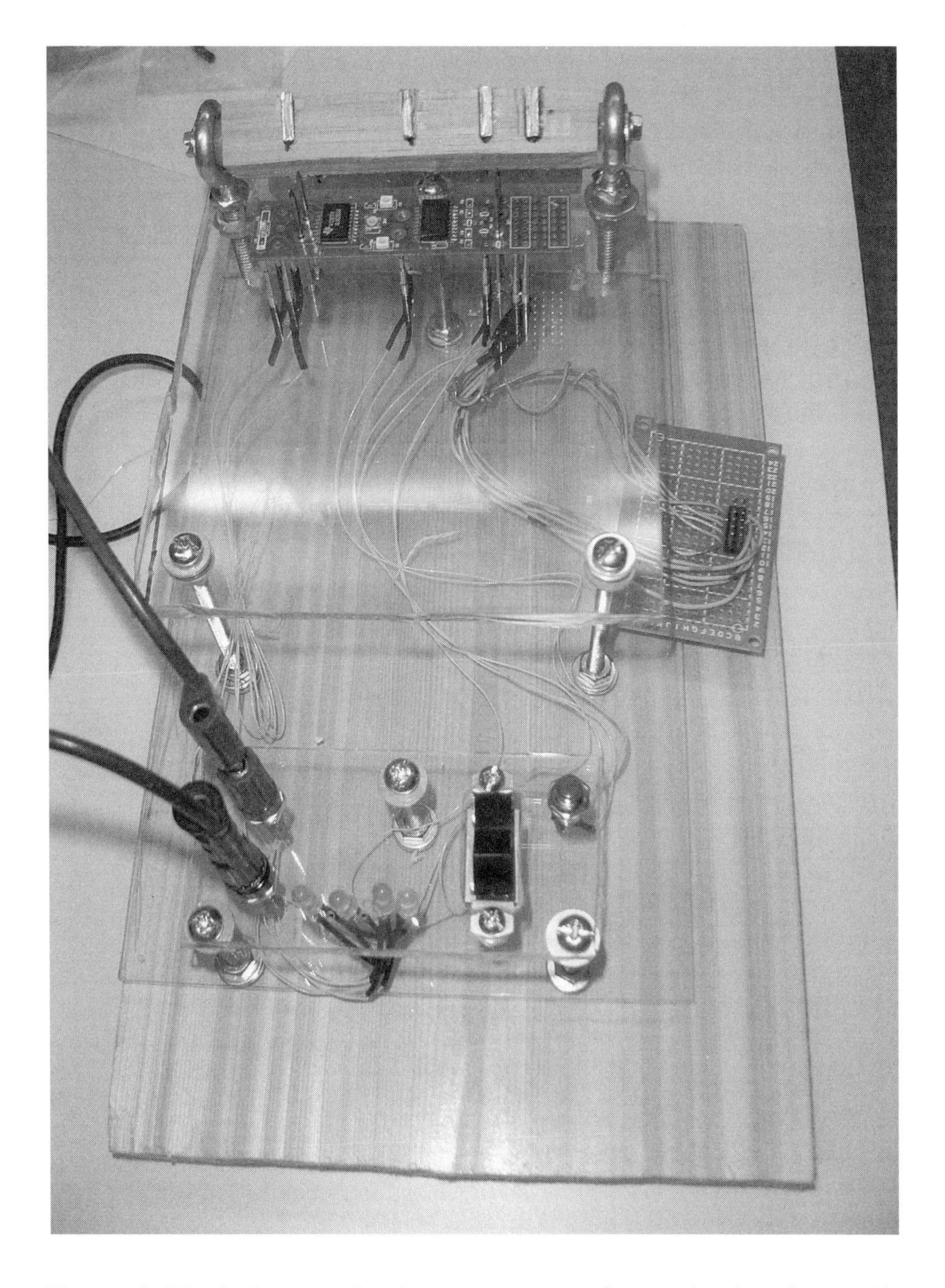

Figure 3.14 A photograph of a prototype test fixture developed to verify manufacturing quality of the Stiquito controller board.

Conclusion

In this chapter, PCB layout and manufacturing were described. In the schematic capture process, the component connections are represented with symbols and lines as well as text to define values and other information. In the layout process, the schematic information is converted to physical information about the components by changing the symbols to footprints. The same symbol on the schematic could result in several different footprints. The board traces and holes are represented with lines, polygons, and circles of various colors to show the relation between the different layers of the PCB. During the manufacturing process, the drawings of the PCB are used to make the real board with the use of masks and stencils that are made directly from the drawings. Different layers are made by various chemical and mechanical processes. Finally, in the assembly process, the PCBs are populated with the components listed in the schematic of the circuit. Automated and manual component placement is made possible by the information provided in the schematic and parts list.

As you can see, collecting and organizing data about the components and connections is important in PCB layout and manufacturing. CAD tools help with maintaining this information, but discipline and practice are needed to make quality PCBs and avoid possible mistakes. In the real world, a lot of interaction between the designers and manufacturers ensures quality products.

References

[1] Conrad, J. M., and J. W. Mills, *Stiquito for Beginners: An Introduction to Robotics,* IEEE Computer Society Press: Los Alamitos, CA, 1999.

[2] Conrad, J. M., and J. W. Mills, *Stiquito: Advanced Experiments with a Simple and Inexpensive Robot,* IEEE Computer Society Press: Los Alamitos, CA, 1997.

Sample Hardware Designs

A design of the Stiquito controller board using the design software ExpressPCB (www.expresspcb.com) can be obtained online at the IEEE CS Press website, http://computer.org/books/Stiquito. Look for the *Stiquito Controlled!* selection. Please note that this design is not the final version of the Stiquito controller board.

Chapter 4

Building Stiquito Controlled

Introduction

Legged robots are typically large, complex, and expensive. These factors have limited their use in research and education. Few laboratories can afford to construct 100-legged robot centipedes, or 100 six-legged robots to study emergent cooperative behavior; few universities can give each student in a robotics class his or her own walking robot.

Stiquito Controlled! Making a Truly Autonomous Robot. By James M. Conrad
ISBN 0-471-48882-8 © 2005 IEEE Computer Society

A small, simple, and inexpensive six-legged robot that addresses these needs is described in this chapter. The robot is 75 millimeters long, 70 millimeters wide, 25 millimeters high, and weighs 10 grams. When a circuit board is mounted on top of the robot, the assembly extends another 10 millimeters lengthwise. The robot is constructed of fewer than 45 parts of which 12 move: six legs bend in response to six Flexinol® actuator wires. Flexinol® is an alloy of nickel and titanium (nitinol) that contracts when heated. It is also called shape-memory alloy [1, 2]. Most parts of the robot perform more than one electrical or mechanical function, but the design can be easily modified. For example, pairs of legs and actuators can be replicated to produce a mechanical centipede with flexible joints between leg segments. The legs can be actuated individually or in groups to yield tripod, caterpillar, or other gaits. The robot is named Stiquito after its larger and more complex predecessor, Sticky.

The robot is intended for use as a research and educational platform to study computational sensors [3, 4], subsumption architectures [5], neural gait control [6], behavior of social insects [7], and machine vision [8]. The robot may be powered and controlled through a tether, but these instructions show how to assemble the robot to work autonomously with an on-board power supply and attached electronics. It is capable of carrying up to 50 grams while walking at a speed of 3 to 10 centimeters per minute over slightly textured surfaces such as pressboard, short-pile indoor–outdoor carpet, or poured concrete. The feet can be modified to walk on other surfaces. The robot walks when heat-activated Flexinol® actuator wires attached to the legs contract. The heat is generated by passing an electric current through the Flexinol® wire (see Figure 4.1).

The Flexinol® wire translates the heat induced by an electric current into mechanical motion, replacing stepping motors, screws, and other components otherwise needed to make a leg move. The mechanical motion results from changes in the crystalline structure of Flexinol®. The crystalline structure is in a deformable state (the martensite) below the martensite transformation temperature, M_t. In this state, the wire may change its

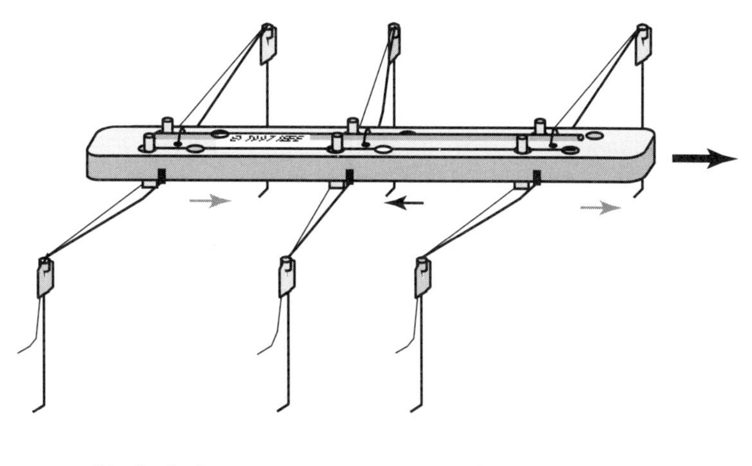

← Flexinol wire actuator contracts, causing ratchet foot to catch surface as leg bends backward

→ other legs slide forward

➞ robot moves forward

Figure 4.1 How the Stiquito robot walks (controller board not attached for clarity of drawing).

length by as much as 10%. The Flexinol® wire provided is an expanded martensite (that is, a *trained* wire).

When the wire is heated above the austenite transformation temperature A_t (1 in Figure 4.2), the crystalline structure changes to a strong and undeformable state (the austenite). As long as the temperature of the wire is kept slightly above A_t, the wire will remain contracted. During normal use of the Flexinol® wire, a recovery force, or tension, is applied while it is an austenite.

When the temperature falls below M_t, the austenite transforms back into the deformable martensite (2 in Figure 4.2), and the recovery force pulls the wire back into its original, expanded form. If no recovery force is applied as the temperature falls below M_t, then the wire will remain short as it returns to the martensite (3 in Figure 4.2), although it can recover its original length by cycling again while applying a recovery force. If the wire is heated too far above A_t, then a new, shorter length results upon transformation to the martensite; the "memory" of the original, longer length cannot be restored.

Flexinol® wire will operate for millions of cycles if it is not overheated and if a suitable recovery force is applied during each transformation. Stiquito's controller prevents overheating if used as directed. Autonomous controllers and software must limit the current supplied to the Flexinol® actuator wires to avoid overheating them. The music-wire legs provide the correct recovery force.

The actuators, legs, and power bus combine to route power, provide the recovery force, and support the robot (see Figure 4.3).

Preparing To Build Stiquito

The robot kit that is included in this book is a slight departure from the kit in the other Stiquito books, *Stiquito: Advanced Experiments* [9] and *Stiquito for Beginners* [10]. The kit in the previous books was typically assembled without screws and included a manual controller. One would make these robots walk by applying voltage to the legs via the simple switches on the manual controller. Although

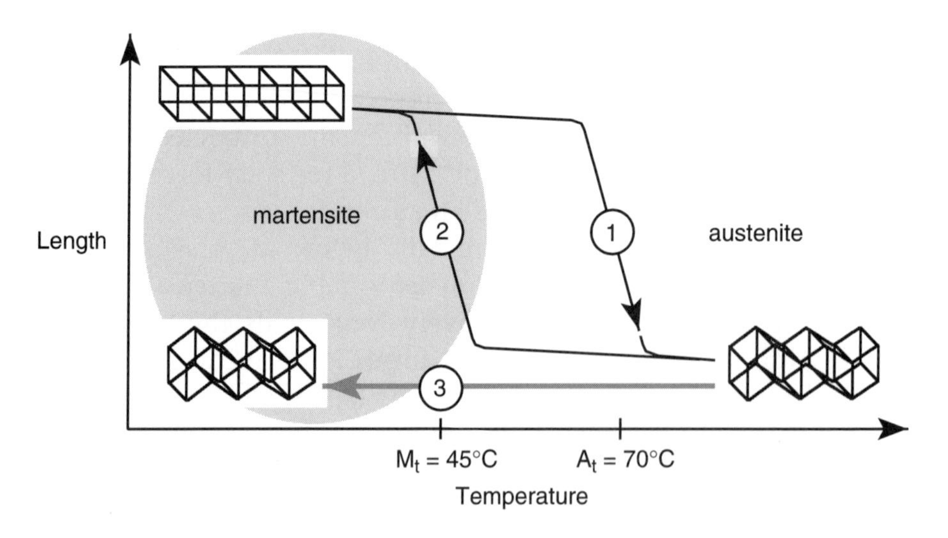

Figure 4.2 Changes in crystalline structure of Flexinol® with temperature.

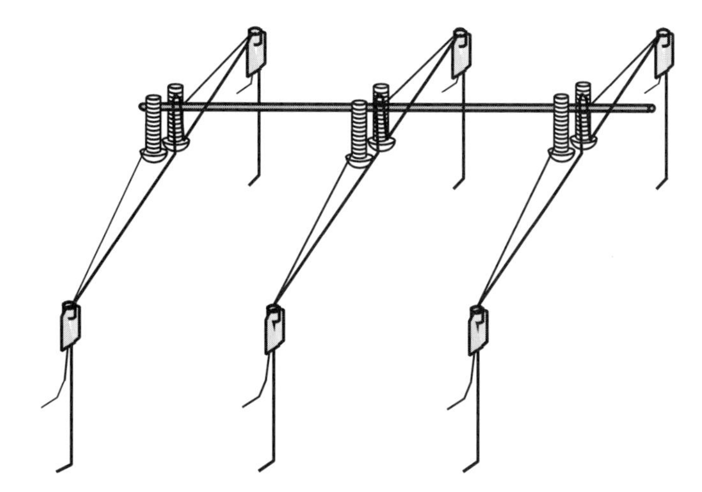

Figure 4.3 Actuators, legs, and power bus.

the "Stiquito Controlled" kit is more complex because of the included microcontroller, its assembly has been simplified. There are fewer parts to cut and crimp. The time required to assemble "Stiquito Controlled" is one to two hours, compared to up to six hours with previous kits.

All parts and tools must be on hand before building the robot. This book contains all of the materials needed to construct Stiquito. Check the materials listed in the parts list (Table 4.1) against those in your kit. If anything is missing, contact missingparts@stiquito.com. Tools are not supplied with the kit. The tools needed to construct the robot are typically available in electronics supply stores or hobby shops, and are listed in Table 4.2.

PRECAUTIONS

- Always follow the manufacturer's instructions when using a tool.
- Wear safety glasses to avoid injury to your eyes from broken tools or pieces of metal that may fly away at high velocity as a result of bending or cutting. Be especially careful when bending or cutting music wire with wire cutters (the cut wire can be forced into your finger, for example).

Table 4.1 Parts list for the Stiquito controlled robot

Photo	Amount	Part number	Item
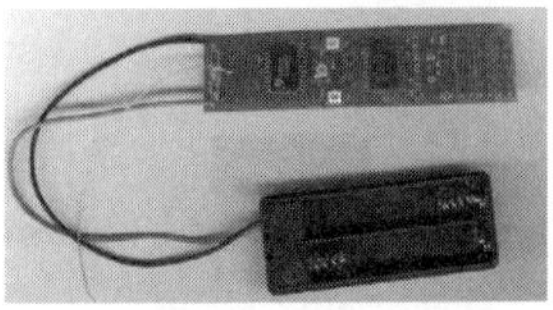	1	ST-202	Stiquito Controller Board with attached two-cell AAA battery holder and 30 AWG copper wire-wrap wire
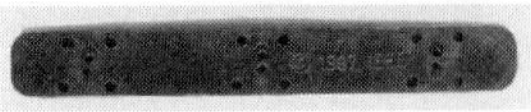	1	ST-100	Molded plastic Stiquito Body
	1 × 100 mm	K&S Eng. No. 100	1/16″ outside diameter aluminum tubing
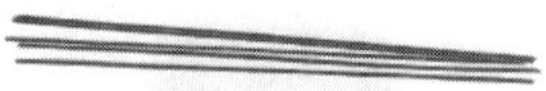	5 × 100 mm	K&S Eng. No. 499	0.020″ music wire
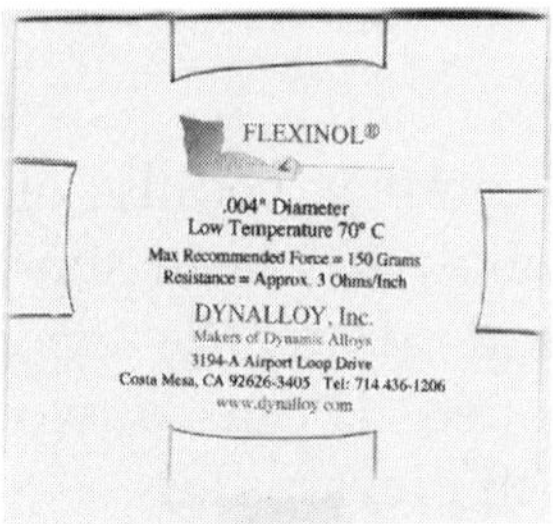	600 mm	Dynalloy 0.004″ Dia 70 C	0.004″ (100 μm) Flexinol® wire
	70 mm	Generic	20 AWG copper hook-up wire (bus bar)
	10 mm	K&S Eng. No. 101	3/32″ outside-diameter aluminum tubing
	1 each	Generic	320 or 360 grit sandpaper
	1 each	Generic	600 grit sandpaper
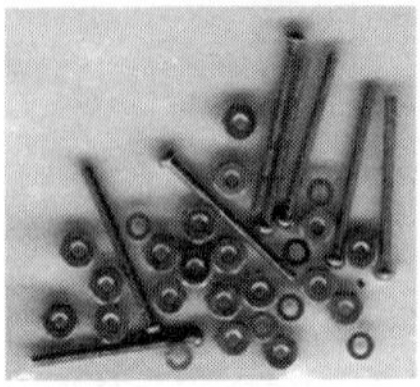	7 each	Generic	Brass screws, 0-80, 1/2″ long (including one spare)
	19 each	Generic	Brass nuts, 0-80 (including one spare)
	7 each	Generic	Brass washers, size 0 (including one spare)

Table 4.2 Tools needed to build the Stiquito controlled robot

Photo	Item
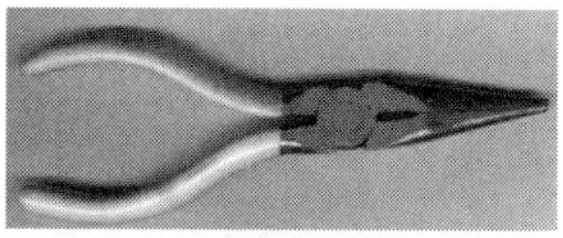	Needle-nose pliers
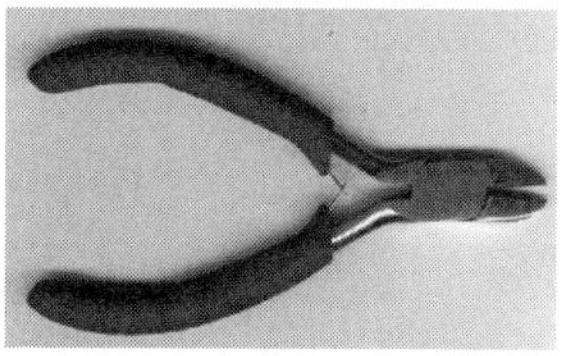	Wire cutters or sharp scissors
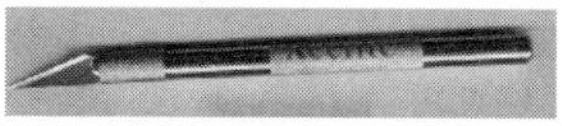	Small knife (X-Acto type)
	Ruler graded in millimeters
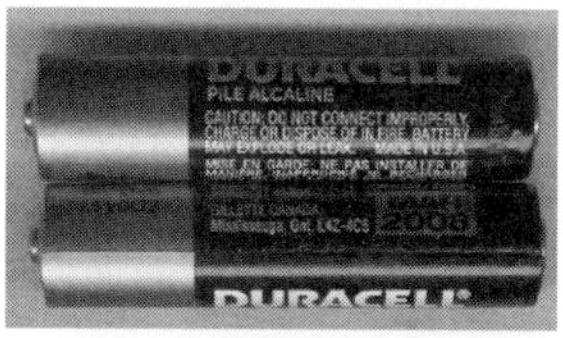	Two AAA batteries
	Volt–ohm meter (or use two AAA batteries and holder)
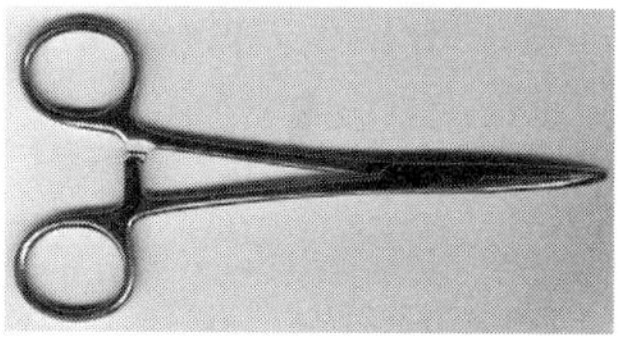	Hemostat (optional)

- Use a piece of pressboard, dense cardboard, a cutting board, and so on to protect your work surface, if necessary.

Before You Build: Required Assembly Skills

A clear workspace and a relaxed frame of mind will be helpful during construction, especially when installing the Flexinol® actuators. Correct installation of the actuators will result in a robot that walks well, whereas a sloppy job will almost certainly lead to one that barely twitches.

Please remember that assembling Stiquito requires hobby-building skills. This section has been included for the benefit of those readers who have not assembled kits before. Practice the skills needed to build Stiquito before assembling the robot, using scrap plastic and thin wire. The kit has extra material in case of mistakes, but there is not enough with which to practice.

Soldering. To build the basic Stiquito robot, you *do not* need to solder any parts. The Stiquito controller board has been preprogrammed to make your Stiquito robot walk with one degree of freedom. Although this board does not include the materials needed to reprogram the microcontroller, several features were added so that users could solder low-cost connectors to the board and expand its capabilities. If you implement any of this functionality, you must solder additional connectors/jumpers to the board before you proceed. Go to the end of this chapter for specific soldering instructions.

Measuring. Following the adage "measure twice, cut once" will prevent most mistakes. Use any metric rule graded in millimeters.

Cutting. Before cutting, check that your fingers are not in the way of your knife, and that a slip of the knife will not damage anything nearby. Direct the knife away from yourself to avoid injury. Make small cuts to avoid removing too much material or making too large or deep a cut.

Deburring. Cutting may leave rough edges (burrs) on some parts. Remove the burrs by sanding the rough edge or trimming the burr with a small knife. Leaving burrs on parts, especially crimps, may cause the Flexinol® actuators to break.

Sanding. The ends of aluminum tubing should be sanded with fine (320 or 360 grit) sandpaper to deburr them. Lightly sand Flexinol® and music wire with ultra-fine (600 grit) sandpaper to remove oxide. Sand the wire after it is bent or knotted to avoid breaking it; sanding wire too much may weaken it enough to break during assembly or when the robot is operating.

Knotting and crimping Flexinol® Wire. Flexinol® is similar to stainless steel. The 0.004″ Flexinol® used in this robot can be knotted without breaking the wire, as long as the knot is not tightened excessively. Knotting and crimping Flexinol® wire is the most reliable way tested to attach the actuators. Flexinol® actuators must be taut and attached so they cannot pull loose if this robot is to walk well. The knot-and-crimp attachments have proven reliable for over 300,000 cycles (approximately 100 hours of continuous walking).

There are other ways to anchor Flexinol® actuators to the legs, but none are as effective as crimps. A U-shaped bend in the Flexinol® wire can pull far enough out of a crimp to reduce leg motion. Soldering is difficult to control because the wire contracts and may lose its "memory," and soldering and epoxying Flexinol® wire may not hold under repeated actuation. Pinning or screwing the Flexinol® wire to the leg is more complex than knotting and crimping (but screws are easy to use on the body).

To tie a knot (Figure 4.4), make a loop in the wire, run one end of the wire through the loop to make an overhand knot, then pull by hand to decrease the diameter of

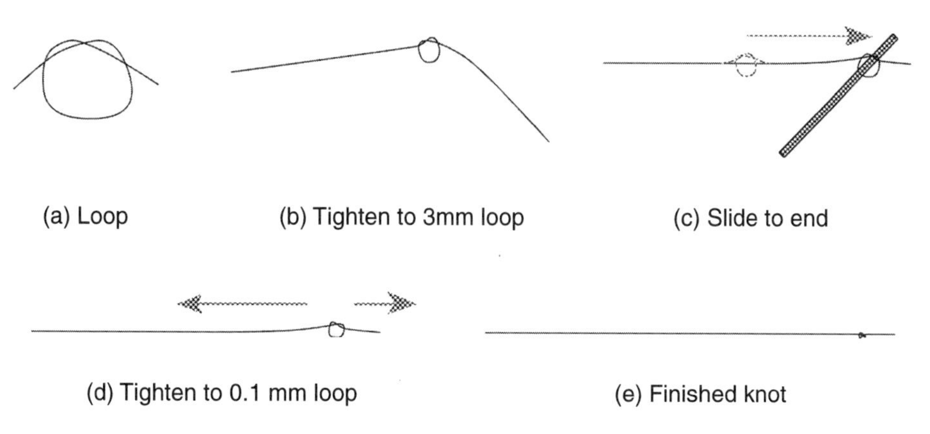

Figure 4.4 Knotting Flexinol®.

the knot's loop to about 3 mm. Slide the knot nearly to the end of the wire using a length of stiff wire, then grasp the end of the wire nearest the knot with the needle-nose pliers, and, holding the other end of the wire (or the crimp if one is attached to the other end) with your hand, pull sharply several times with the needle-nose pliers to tighten the knot. The knot is tight enough when a small loop, approximately 0.1 mm in diameter, remains. Continuing to tighten the knot may break the wire.

Crimps (Figure 4.5) are hollow connectors that are squeezed shut to hold, attach, or connect one or more objects. This robot uses short lengths of aluminum tubing as crimps to anchor the knotted Flexinol® actuator wires securely. Leg crimps hold both the actuator wire and the music-wire leg. Leg crimps directly attach and electrically connect the actuator wire to the music-wire leg. We use screws to attach the Flexinol® to the body.

Fixing Mistakes

No matter how carefully you work, mistakes do happen. Most are easy to fix, because almost all steps in the construction of Stiquito allow some tolerance, except for ten-

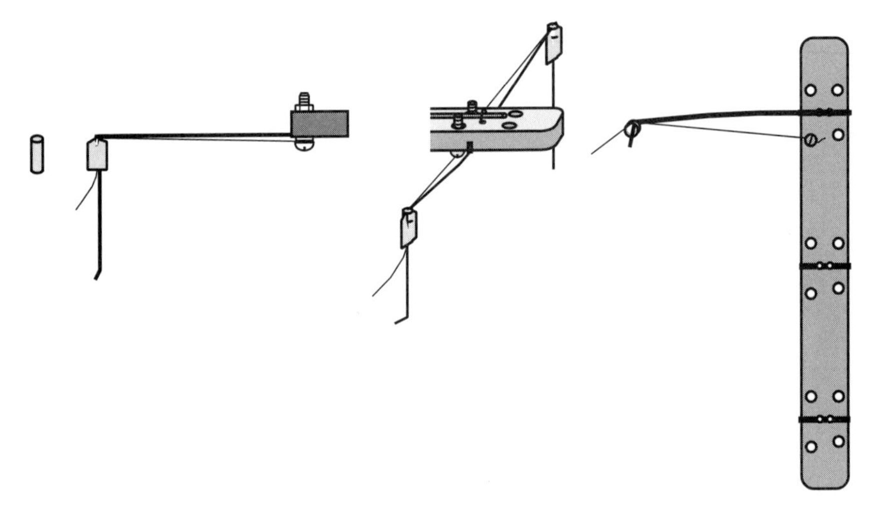

Figure 4.5 Leg crimp (views with repsect to leg crimp).

sioning the Flexinol® actuator wires, where no slack is allowable. Here are some common problems, and how to work around them:

- *Music wire bent incorrectly.* 90° bends or greater can be rebent gently once or twice before breaking the wire. Bends less than 90°, such as the 15° V-clamp in the legs, usually break if rebent.
- *Knot in wrong place.* Tie a new knot and keep going. You may want to cut the Flexinol® to 70 mm lengths instead of 60 mm lengths for this reason. Untying tight knots in Flexinol® usually breaks it.
- *Crimp must be removed or replaced.* Crimps can be gently squeezed across the wide dimension to undo them, but should then be discarded. Extra tubing is provided to make new crimps.

Constructing Stiquito

Stiquito has four major assemblies (Figure 4.6): the controller printed circuit board (PCB), the body, the legs and power bus, and the actuators. The actuators are made of Flexinol® wire.

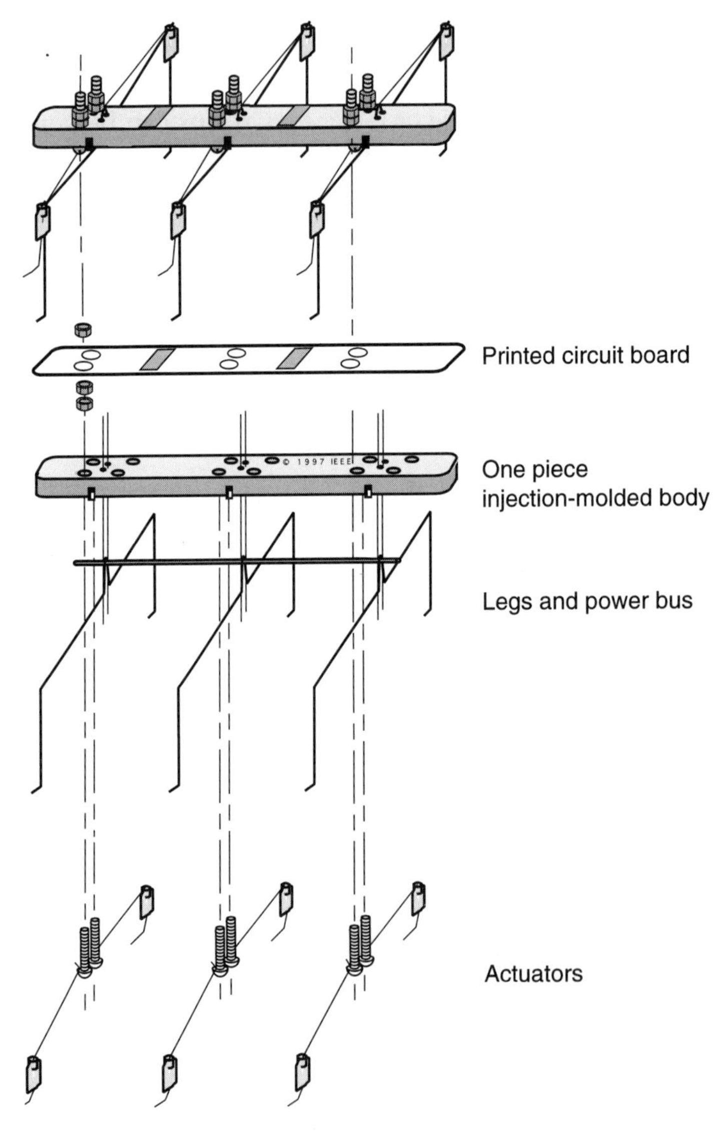

Figure 4.6 Stiquito assemblies.

The Body

The body (Figure 4.7) provides structural strength and locates the attachment points for the legs and the Flexinol® actuator wires. The body is molded with holes and grooves. Examine the plastic to ensure that every hole goes all the way through and that there are no rough edges.

The plastic body has 18 holes. The smallest set of holes is used to attach the legs to the body. The set of six large holes that have three holes slightly offset will be used to assemble the leg actuators. The following instructions use only these twelve holes.

The other set of six holes that are parallel to the small holes can be used for other purposes, including assembling Stiquito with legs that have two degrees of freedom (see Chapters 6 and 7).

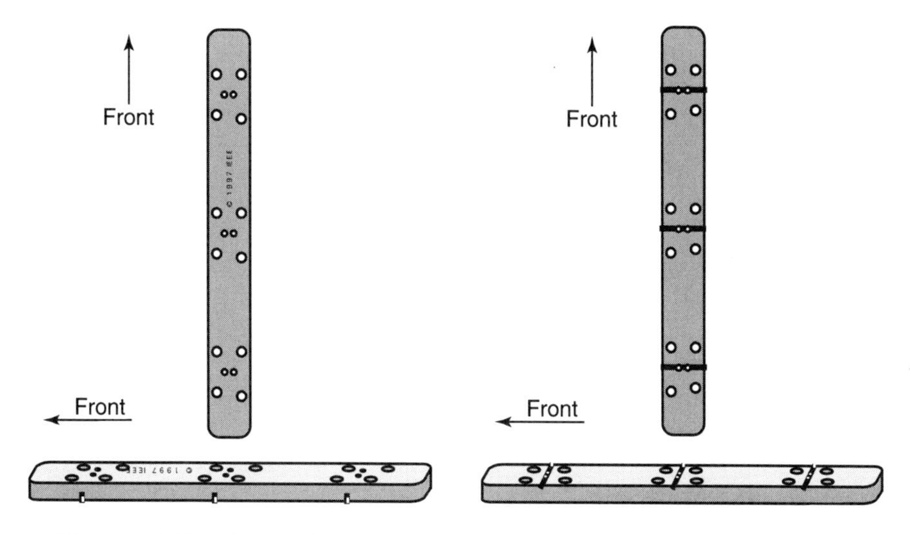

Figure 4.7 The molded body. Left—top side. Right—bottom side.

The Legs and Power Bus

The legs (Figures 4.8 and 4.9) are assembled in pairs from three 100 mm lengths of 0.020″ music wire. The music wire legs perform three functions:

1. *Support.* The legs support the weight of Stiquito and its battery and control electronics. Because the wire is bent to fit into a leg-clip groove molded in the body, each leg in the pair is mechanically isolated.
2. *Power distribution.* All legs share a common electrical power connection to the power bus that routes current to the Flexinol® actuator wires. The V-bend in the music wire clamps the power bus to the top of the body and electrically connects it to the legs.
3. *Recovery force.* The music wire acts as a leaf spring to provide recovery force for the Flexinol® wire actuator. Without this spring, or if the actuator is attached loosely, the Flexinol® wire will contract, but fail to return to its original extended length.

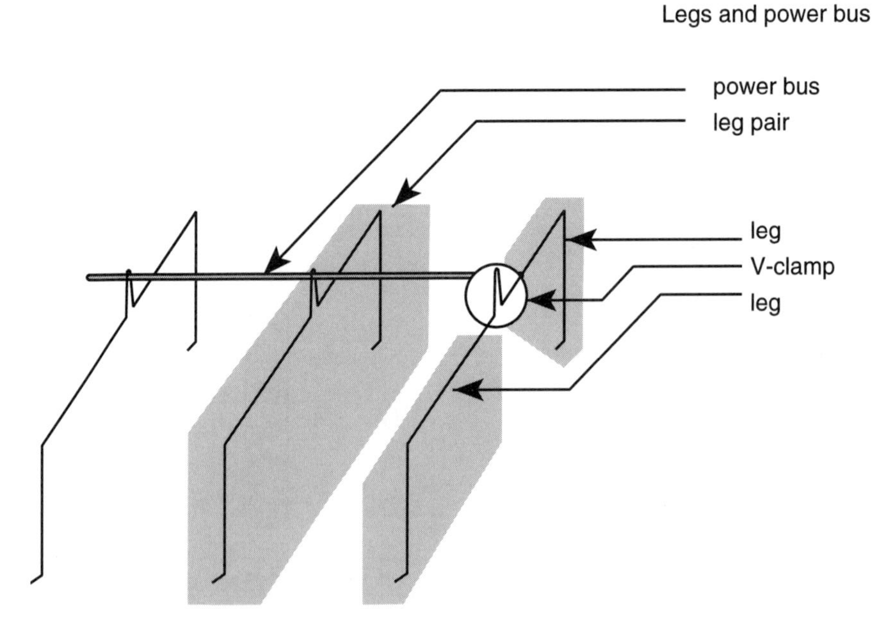

Figure 4.8 Legs and power bus.

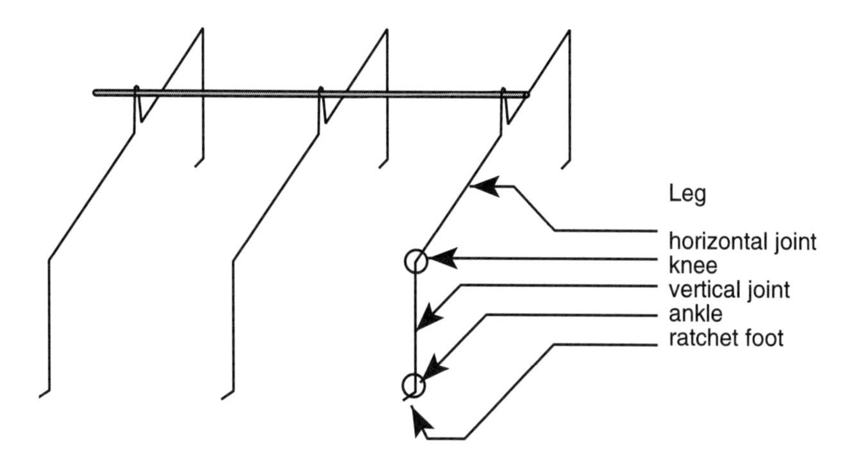

Figure 4.9 Leg detail.

Begin assembling the legs by using three 100 millimeter lengths of 0.020″ music wire. Bend each music wire in the middle to a 15° angle (Figure 4.10). Do not bend the wires too far, or they may crack or break. The apex that forms the V-clamp should be rounded, not sharp. Lightly sand the inside of each V-clamp with the 600 grit sandpaper to remove oxide.

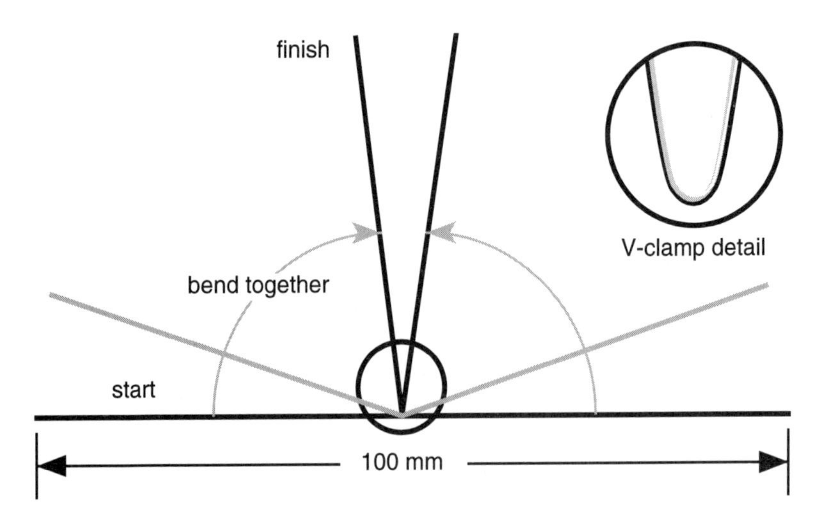

Figure 4.10 Bending the legs (enlargement 6 ×).

Remove the 70-millimeter length of 20 AWG copper wire from the kit. This is the power bus (Figure 4.11). Lay the power bus along the top of the body between the leg holes. Leg holes are the smallest holes on the body. Temporarily clamp the power bus by bending the legs together, inserting them through the leg holes, and then pulling the legs through from the other side until the power bus is held tightly by the V-clamp (Figure 4.12).

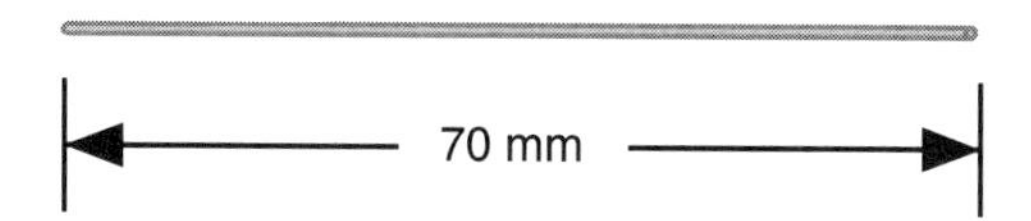

Figure 4.11 Power bus.

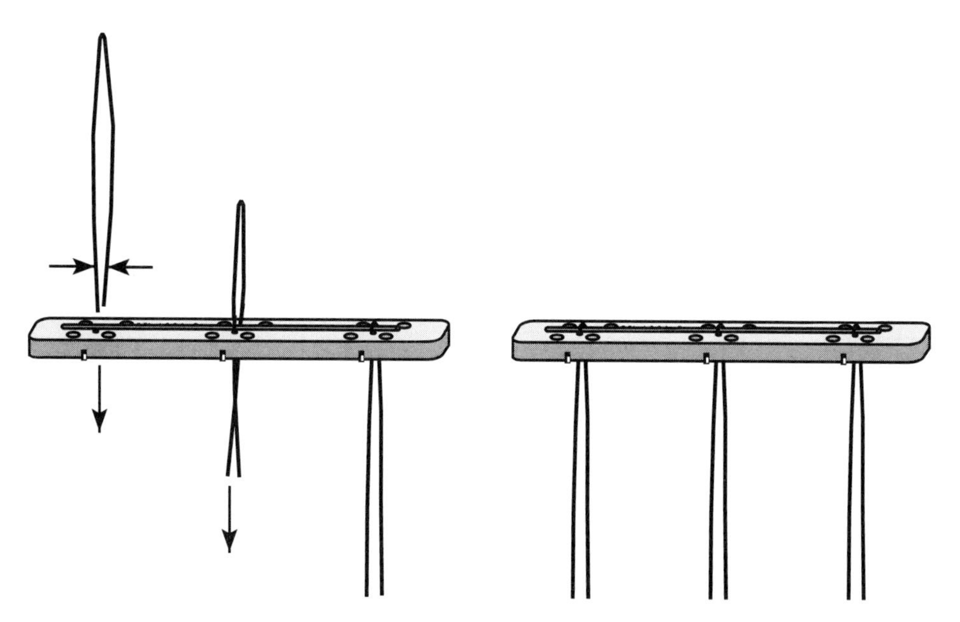

Figure 4.12 Temporarily clamping power bus with the legs.

Turn the body over and permanently clamp the power bus, simultaneously attaching the legs by spreading each leg in the pair outward by hand, while at the same time pulling upward on the legs (Figure 4.13). When the legs are almost horizontal, grasp each leg in turn with the needle-nose pliers and firmly bend it downward, while continuing to pull outward, until the leg snaps into the clip groove.

At this point, the power bus should be securely clamped into place. Check to ensure that it will not touch any of the body screws (insert a screw and check with the volt–ohm meter). Also, check that the electrical connection between the power bus and the legs is good. It should be less than 2 ohms.

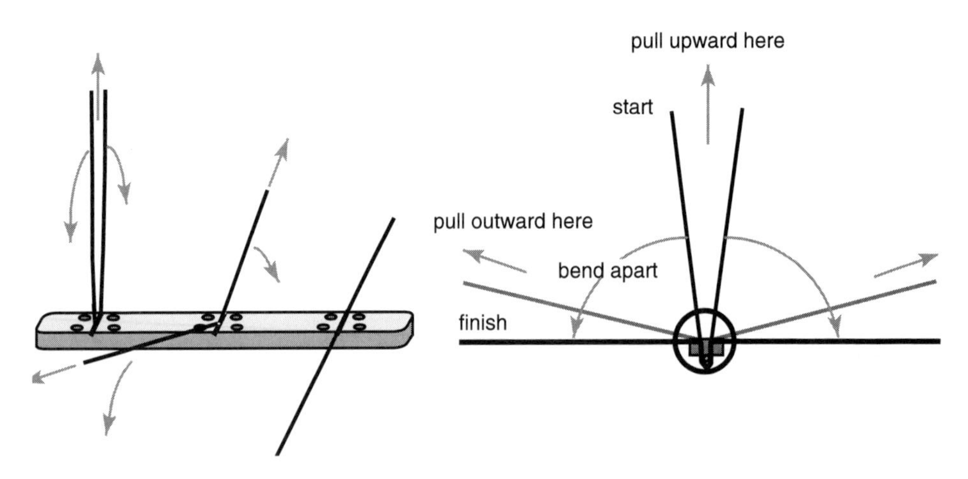

Figure 4.13 Permanently attaching the legs and clamping the power bus.

Adjust the legs so that they are in a plane horizontal with the bottom of the body and parallel to each other. Working with the bottom of the body facing upward, form the knee, which separates the horizontal joint from the vertical joint of each leg, by bending the music wire 90° about 30 mm from the edge of the body (Figure 4.14). Adjust the vertical joints so they are parallel.

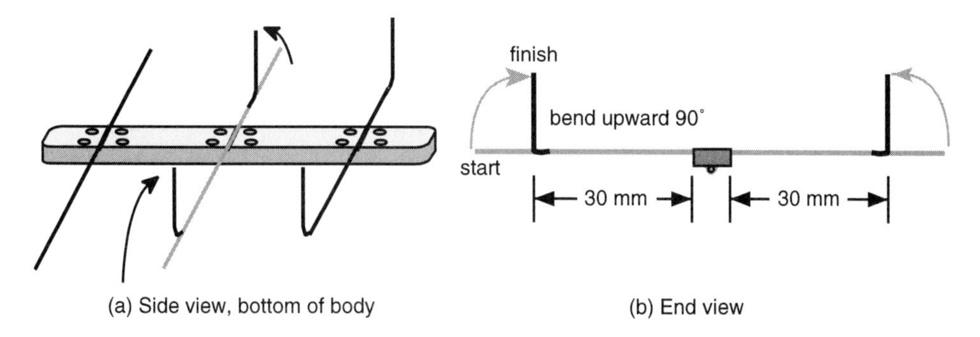

Figure 4.14 Forming the knee and the horizontal and vertical joints.

Trim the vertical joints using the wire cutters so that all legs touch the ground (Figure 4.15). Do not bend the feet yet; you will do that toward the end of the robot assembly. This completes assembly of the legs and power bus.

Do not bend the music wire to make the ratchet feet at this time. Wait until the actuators have been completed and tested.

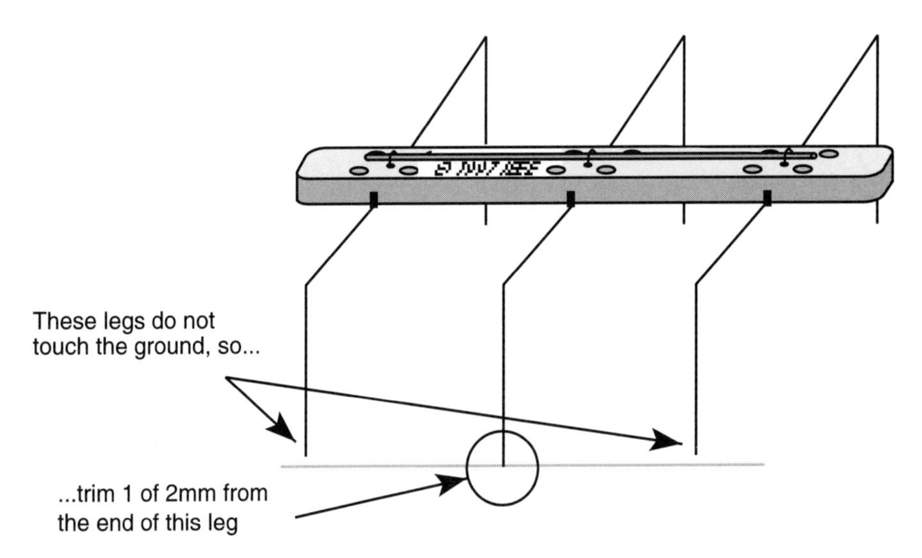

Figure 4.15 Trimming the vertical joints.

The Actuators

Stiquito is small and simple because it uses Flexinol® actuator wires.

The following steps are needed to build the actuators (Figure 4.16), attach them to the legs and body, and form the ratchet feet.

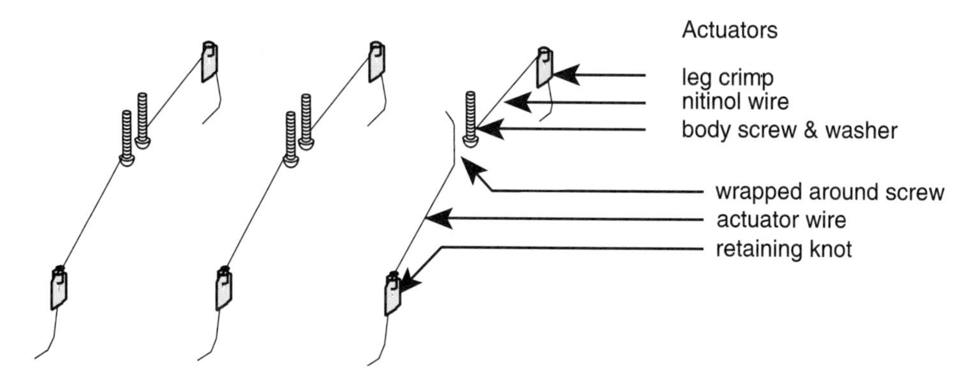

Figure 4.16 Actuators.

Preparing the Screws on the Body. Turn Stiquito upside down with the front of the body pointing away from you. Place one washer on each of six screws (there is one extra screw and washer). Insert the screws with the attached washers into the three sets of offset holes near each pair of legs. Secure the screws loosely to the body by threading one brass nut on each brass screw by hand (Figures 4.17 and 4.18).

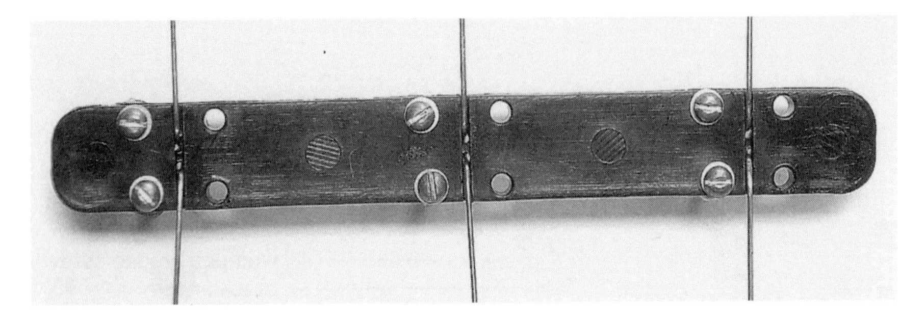

Figure 4.17 Adding six screws, washers, and nuts to the Stiquito body.

Figure 4.18 The specific stack-up of screws, washers, and nuts on the Stiquito body.

To verify that you have inserted the screws in the correct holes, remove the Stiquito controller board from the plastic bag. Place the Stiquito robot on its feet with the front of the robot to your left and the screws to the right of each leg. Holding the board by the edges, orient the board so that the prototype end of the board (with six rows of holes) are to your left and the electronic components are facing up (Figure 4.19). Notice how the very large holes in the board are also offset like the Stiquito body. Align the screws of the Stiquito so that they match the large holes in the board, and *gently* push the body and board together. If the holes do not align, then you probably did not put the screws in the correct holes of the Stiquito plastic body. Remove the screws and start the step over from "Preparing the Screws on the Body" above. Remove the Stiquito controller board to redo this step or to continue.

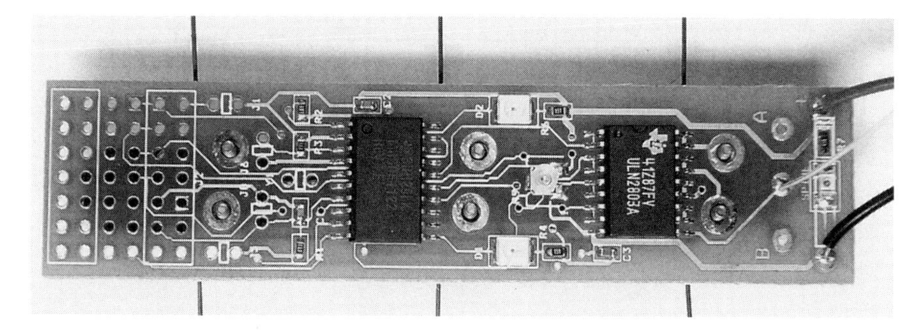

Figure 4.19 Checking alignment of the Stiquito controller board with the Stiquito body.

Make Leg Crimps (and the Power Bus Crimp) (Figure 4.20). Next, select the aluminum tubing from the kit. It will be used to make the leg crimps. Using a knife, cut six 4 mm leg crimps from the aluminum tubing. *Do not* use a wire cutter. A wire cutter will squeeze the ends of the tubing shut. Sand the ends of the crimps, then run the end of the knife through them to deburr the ends.

Cut Flexinol® Wires to Size. Begin making the actuators by cutting six 55 mm lengths of Flexinol® wire (there will be extra Flexinol® wire left over). It might be preferable to use some of the extra Flexinol® in the kit, and cut six 60 mm lengths to make it easier to wrap the Flexinol® around the screws during the tightening of the body screws (Figure 4.21).

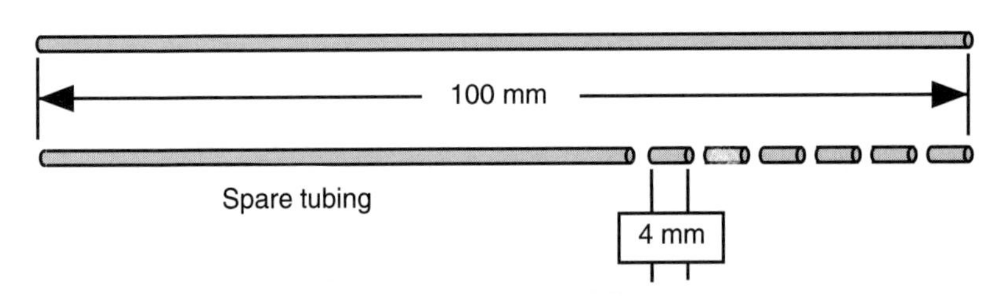

Figure 4.20 Leg crimps.

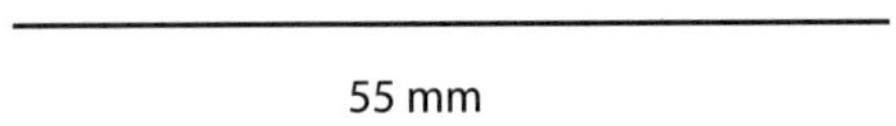

55 mm

Figure 4.21 Flexinol® wire.

Attach Leg Crimps. These steps will have you attach each length of Flexinol® wire to each Stiquito knee using a leg crimp. Tie a retaining knot in one end of the wire. Using the 600 grit sandpaper, *lightly* sand the Flexinol® wire at the knot to remove oxide and improve the electrical connection to the body crimp. To sand, lightly brush the sandpaper across the knot, one stroke on each surface (Figure 4.22).

Take a 4 mm body crimp, insert the nonknotted end of the wire into a leg crimp, and pull it through the crimp; the knot must barely extend out one end (Figure 4.23).

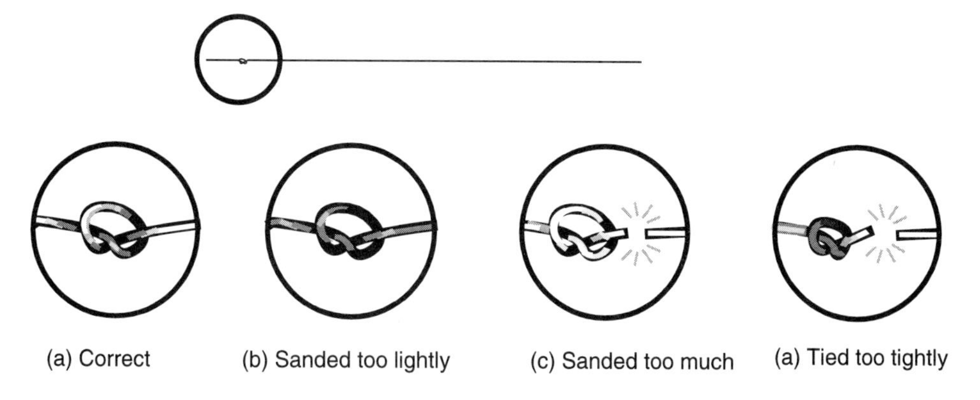

Figure 4.22 Tying and sanding retaining knot (all enlargements × 10).

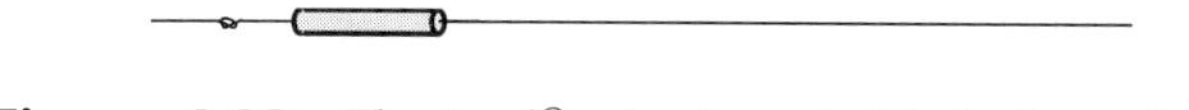

Figure 4.23 Flexinol® wire inserted into leg crimp.

Holding the Stiquito body, slide the leg crimp with the Flexinol® onto the vertical joint of one leg and slide it up to the knee (Figure 4.24). The knot should still extend out of the crimp. Once the tubing is as close to the joint of the leg as possible, slowly pull the wire through the tubing until the knot is inside the tubing about one-half way.

Once the knot at the end of the wire is inside the tubing, using pliers, crimp the entire length of tubing to secure it to the leg. Gently tug up on the Flexinol® wire to ensure the crimp is holding the knot, and Flexinol®, inside (Figure 4.25).

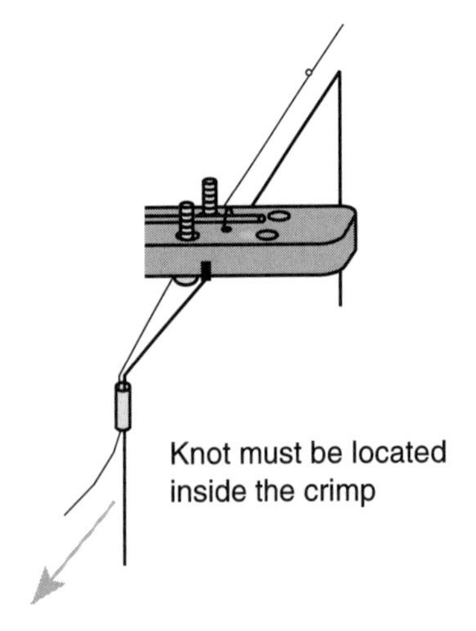

Figure 4.24 Flexinol® wire pulled into leg crimp on the leg.

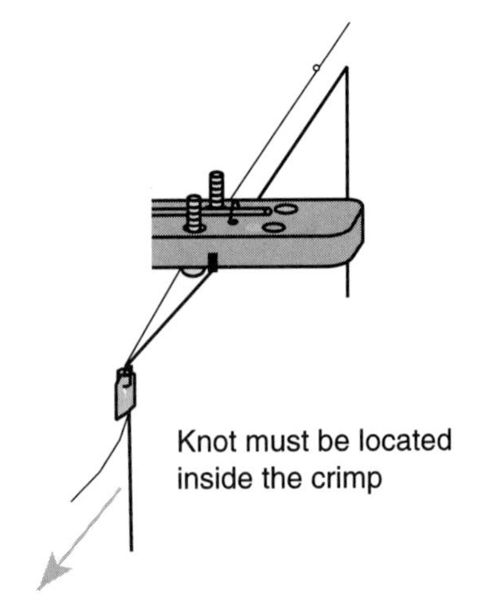

Figure 4.25 Crimp the tubing to hold the knot in place.

Hold the loose end of the Flexinol® wire and pull it toward the Stiquito body screw. Wrap the other end of the Flexinol® wire *clockwise* all the way around the screw once, between the head of the screw and the washer, and pull taut. Be careful not to place the wire between the washer and the plastic body of the Stiquito or proper tension will not be maintained (Figure 4.26).

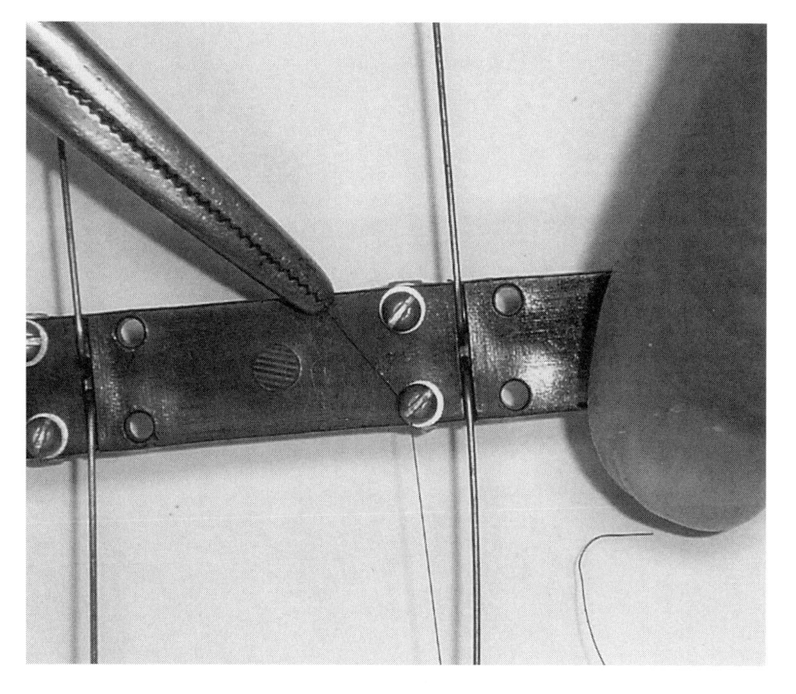

Figure 4.26 Wrap the Flexinol® wire around the screw.

Secure the wire between the head of the screw and the washer. Keep the wire in place by applying pressure with a finger to the head of the screw while finger-tightening the nut on the other side of the body. Once the nut can no longer be tightened by hand, hold the head of the screw in place with the small flathead screwdriver. Continue to turn the nut using the needle-nose pliers until a fair amount of resistance is felt (Figure 4.27). *Caution:* screws can be stripped if nuts are overtightened

Figure 4.27 Tightening the brass nut.

The goal is to have the wire taut so that there is a slight backward bend in the horizontal joint. The slight bend ensures that there is no slack in the Flexinol® actuator wire. There will not be enough tension if there is no bend in the leg. There will be too much tension if the leg is bent more than 2 mm backward at the knee (Figure 4.28). To make the Flexinol® tauter, turn the screw clockwise.

Once the wire is secured, the excess may be trimmed off. It is still a good idea to leave about ¼ inch of wire so that the previous steps can be redone if the tension is not sufficient.

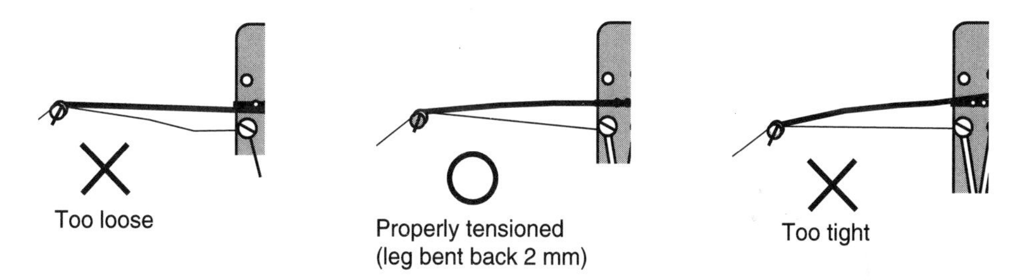

Figure 4.28 Correct tension (O) and incorrect tension (X).

Measure the resistance from the leg near the clip groove to the body crimp where it protrudes above the top of the body. The initial resistance should be between 5 and 7 ohms. The resistance will increase to between 15 and 25 ohms as the connections age. Test the operation of the actuators by applying current from two 1.5 volt AAA cells at the leg near the body and the body crimp for no more than half a second to prevent the actuator wire from overheating (Figure 4.29). The leg should immediately bend backward 3 mm to 7 mm measured at the vertical joint, then return to its original position.

Test the leg and actuator assembly after installing each leg Flexinol® wire. If the test is successful, continue to attach and test the remaining Flexinol® wires. Once all actuator wires are attached, screw a second nut onto each of the screws to "lock" the first nut in place (Figures 4.30 and 4.31).

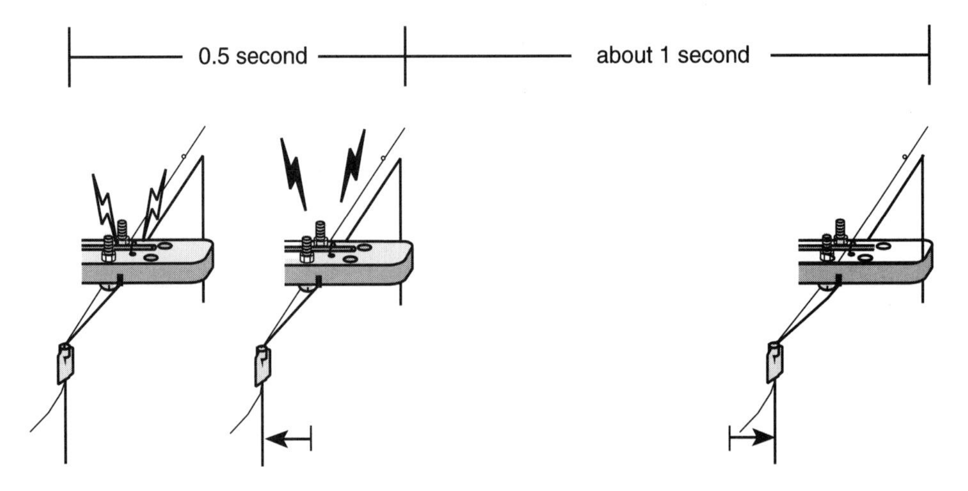

Figure 4.29 Testing the actuator.

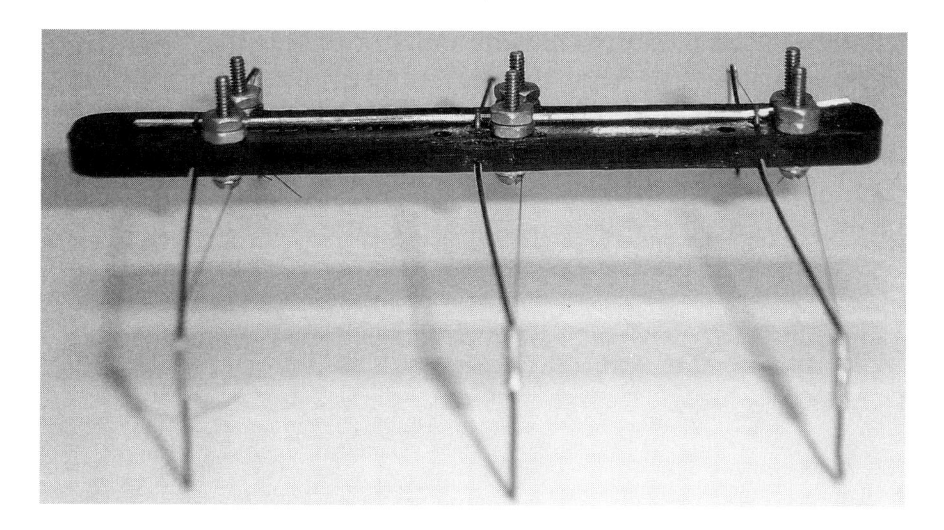

Figure 4.30 Finished actuators—side view.

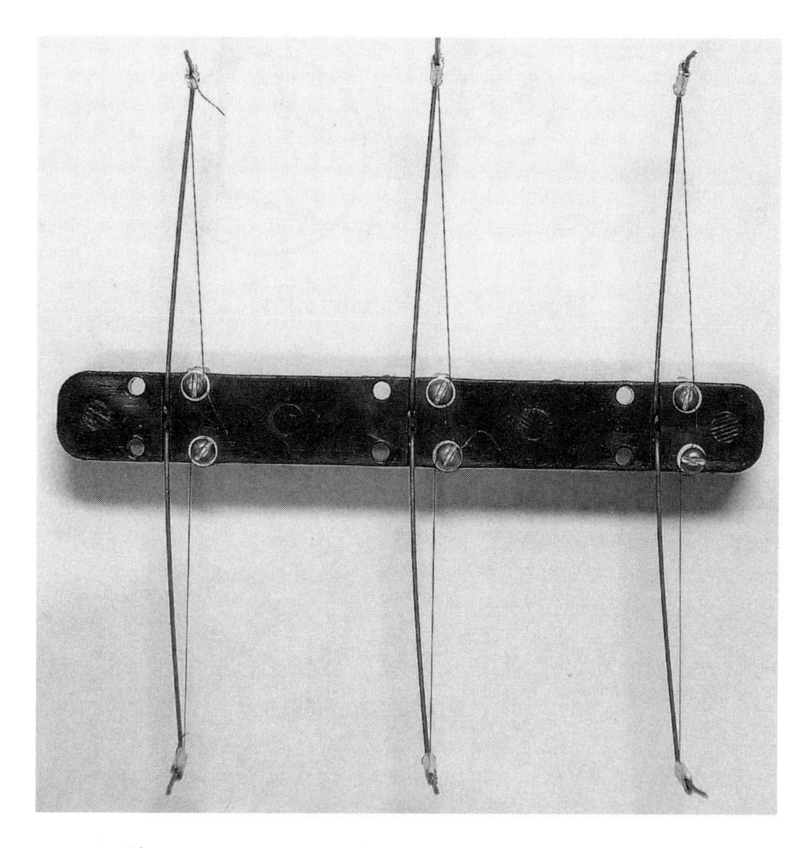

Figure 4.31 Finished actuators—bottom view.

The Ratchet Feet. Be sure to test the actuators before making the ratchet feet in case the leg crimps must be replaced. Form an ankle and ratchet foot by bending the tip of each vertical joint backward and slightly outward. The ratchet foot should be about 2 mm long, and make a 110° angle downward from the vertical joint (Figure 4.32). The ankle faces toward the front of the robot.

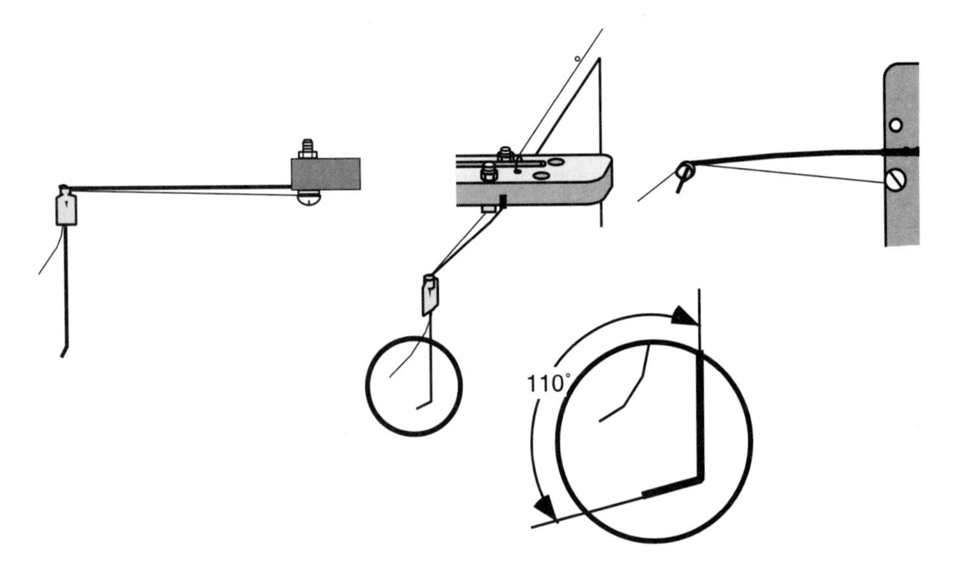

Figure 4.32 Ratchet foot.

The next step to complete your Stiquito robot is to attach the power bus supply wire to the power bus. This wire connects the Stiquito power bus to the circuit board. This wire will be crimped to the power bus to ensure an effective connection.

Cut the 10 mm length of 3⁄32″ aluminum tubing in half. One-half will be used and the other is a spare. Gently lift the back end of the power bus wire and slip a 3⁄32″ aluminum crimp onto the power bus. Remember, the back of the robot will have two of the screws close to the end. Take the stripped end of the 30 AWG wire attached to the printed circuit board and slip it into the crimp as well. Using your needle-nose pliers, squeeze the crimp tight (Figure 4.33).

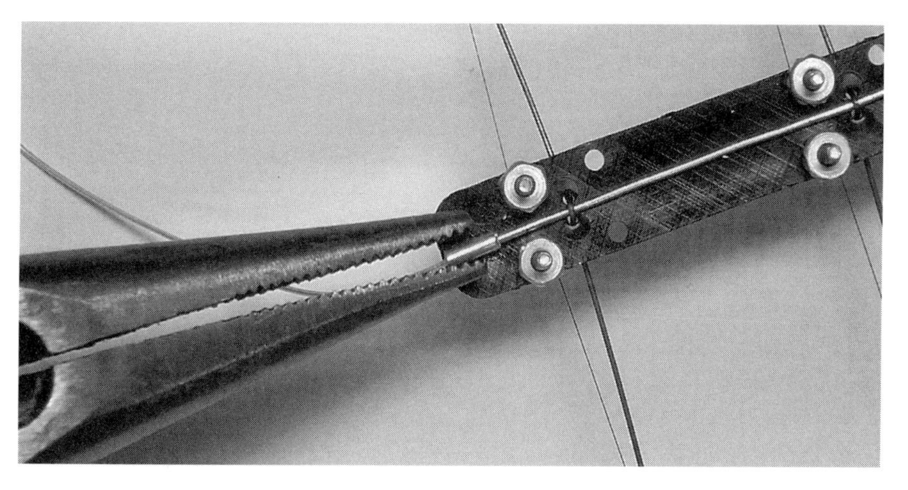

Figure 4.33 Crimping the power bus supply wire to power bus.

Take the Stiquito controller board and loosely place it on the Stiquito robot in the correct orientation, as described earlier. The prototype area of the board should be toward the front of the robot. Push the Stiquito controller board all the way onto the Stiquito robot body screws. Put a brass nut on each of the screws and tighten them until there is some resistance against the controller board (Figure 4.34).

Check that attaching the board to the robot did not cause any of the Flexinol® actuators to loosen. If an actuator did loosen, use a screwdriver to turn the screw clockwise and make the Flexinol® wire taut.

The assembly of your robot is now complete.

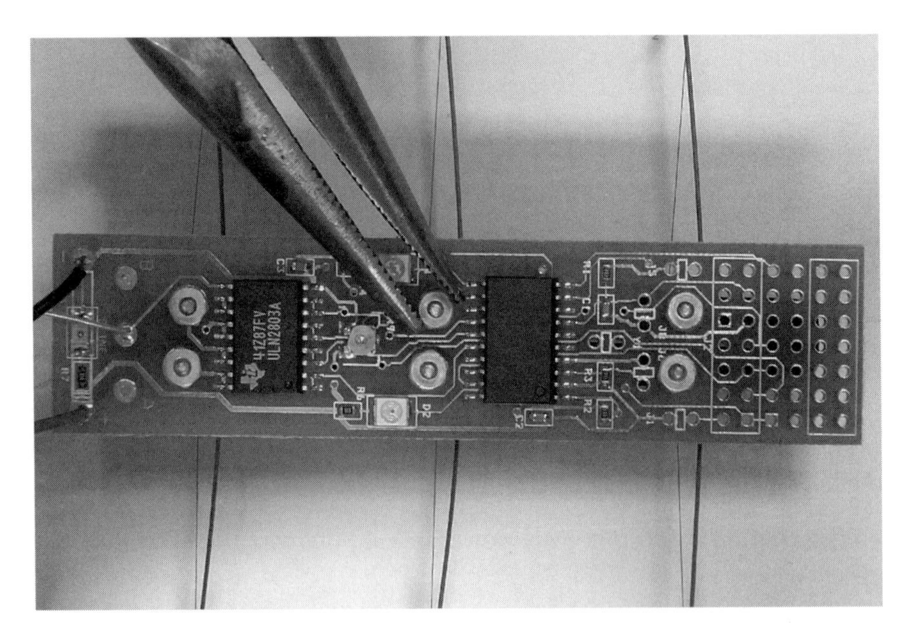

Figure 4.34 Securing the controller board to the robot.

Operating the Stiquito Robot

In order to operate the robot, you will need fresh AAA batteries. Since the board does not have an on/off switch, make sure to remove the batteries from the AAA battery holder when not in use.

Insert the AAA batteries into the battery holder. You should see the power LED light (on the back of the controller board). After a few seconds, you should see the LED on each side of the controller board alternately light. At the same time, the two tripods of the robot should alternately contract.

In order for the robot to walk effectively, three of the legs in one tripod should fully contract backward, then relax and return to their original position; and then the legs in the other tripod should fully contract backward, then relax and return to their original position. If the legs are cycling too fast (they do not fully contract), then you will need to adjust the speed potentiometer, located in the center of the robot, to make the cycle longer. If the legs are staying contracted too long, then you will need to adjust the potentiometer to make the cycle shorter.

To adjust the potentiometer, take a small screwdriver (flat-head or Phillips), place the tip inside the potentiometer (Figure 4.35), and:

- Turn clockwise to make the cycle slower
- Turn counterclockwise to make the cycle faster

Once the legs are contracting at the correct speed, you can attach the battery to the robot using a rubber band. Place Stiquito (Figure 4.36) on a slightly rough surface, and watch it walk!

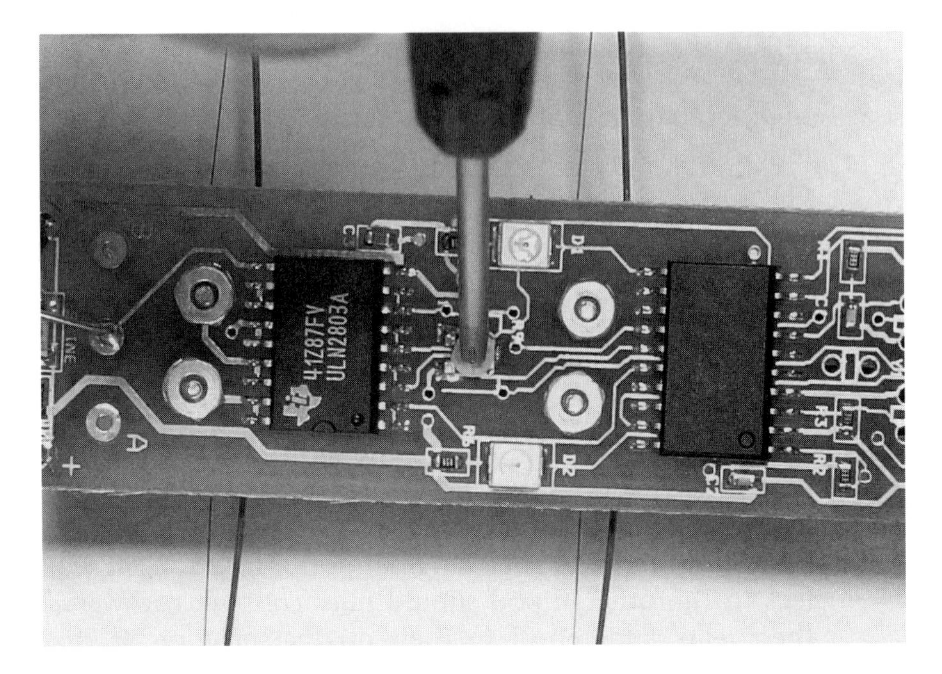

Figure 4.35 Adjusting the speed of Flexinol® wire actuation.

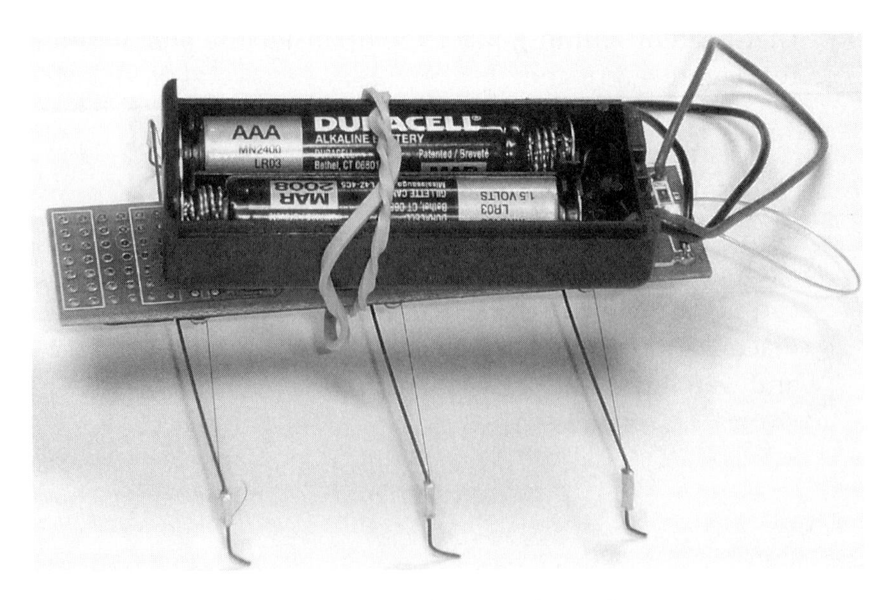

Figure 4.36 The complete robot.

Troubleshooting

Troubleshooting is applied logical deduction. To avoid the frustration encountered when a project fails to work as expected, *expect it not to work.* Take this attitude from the start; think about factors that could affect the operation of the robot, and the behavior (or lack of it) that would result. Then, when the inevitable happens, you will have a set of hypotheses about why the robot failed to work. The hypotheses may be wrong, but that is OK: wrong assumptions lead to right deductions if you are willing to discard assumptions that are not supported by experiment. Simply remember that nothing robotic works until you make it work. When software, hardware, and microcontroller development tools combine to make a robot, you will quickly learn that at least some expertise in several fields are required to design a robotic product.

Stiquito Walks in a Straight Line. This is perfect. You do not have any trouble. Congratulations on a good job!

Stiquito Walks to the Left or Right, But Not Straight. This is OK, and is very common. The reason is that one or more of the legs is crooked, or does not operate at full extension. Check for loose actuator wires, loose crimps, poor electrical connections (look for open or high-resistance paths from the power supply to the legs and back), a weak power supply (Flexinol® draws about 180 mA per leg, which will drain two 1.5 volt cells after several weeks of occasional use); legs that are not parallel, or ratchet feet bent at different angles.

All LEDs are Lit—No Flashing. The microcontroller did not start correctly. Remove then reinsert one battery in the battery holder.

Stiquito Does Not Walk at All. If there is nothing obviously broken or loose (actuator wires, crimps, brass screws), then check the power source. The two 1.5 volt cells may be dead. Check the power bus and the body screws for shorts. If you did not bend the ratchet feet, or if you operate Stiquito on a smooth surface, the legs may move but will not catch the surface, and Stiquito will thrash around but not walk. Stiquito walks best on slightly textured surfaces such as indoor–outdoor carpet, a cloth-covered book, pressboard, or poured concrete. Stiquito walks poorly on glass, smooth plastic, or tiled floors.

Leg Moves to Full Extension (4–5 mm). This is perfect. You obviously built Stiquito carefully.

Leg Does Not Move at All. The probable causes include a very loose actuator wire or dead 1.5 volt batteries. Check the cells with a volt–ohm meter. If the power is OK, then check the electrical connection between the bus and the body screw. Test single actuator wires using 3 volts supplied by two 1.5 volt AAA cells in series. If the electrical connections are good, then examine the actuator wire. If it is loose, or the body screw or leg crimps are not tight, then the actuator is too slack to operate. Tighten the actuator wire (you may have to loosen, then tighten, one or more screws/nuts), then test the leg. It should work.

Leg Moves Slightly (1–3 mm). The actuator is probably loose, but is taut enough to take up the slack, then move the leg. Retension the actuator, turning the body screw. Another cause is increased resistance as the crimps age. Aluminum oxide builds up inside the crimps. Its effects can be alleviated by operating the leg. This causes the retaining knots to expand and improve contact with the aluminum inside the crimp. Squeezing the crimp also helps to improve the electrical connection.

Leg Moves in One or More Jerks. There is probably an intermittent open or shorted connection. If the leg jerks backward continuously in small increments, remove power immediately or the actuator wire may be damaged. Check the assembly for shorts.

Leg Heats Up, Smokes, and/or Melts Plastic Near Body Screws. If you see of smoke, or if you smell something "hot" like burning or melting plastic, then remove power immediately or the actuator wire may be damaged. There may be a short in the wiring. You may be powering a slack actuator for longer than one second. A slack actuator will not work. Continuing to apply power to the leg will only cause the Flexinol® actuator wire to overheat, damaging the wire and heating the body screw enough to melt the plastic body.

Leg Works OK for Awhile, Then Movement Stops Altogether (and the Controller Board Lights are Still Flashing). The actuator wire has developed slack, a connection has broken, or the batteries have lost enough voltage to power the circuit, but not the legs. If the actuator wire is slack, it may need to be retensioned. Try fresh 1.5 volt cells.

Leg Works OK for Awhile, Then Movement Diminishes by 2 mm or More. The actuator is probably somewhat loose, but is taut enough to take up the slack, and then move the leg. Another cause is increased resistance as the crimps age. Apply power to the leg, squeeze the leg crimps to improve the electrical connection, and/or turn the body screws.

Actuator Wire Breaks. If the actuator wire breaks during assembly, then it was sanded too much, or nicked (probably while removing the knot from a music wire leg). Actuator wires may also break from these causes during operation.

The Future of Your Stiquito Robot

Now that you have finished building your Stiquito Controlled robot and it works, you can think about what else you want Stiquito to do. Do you want to make it walk faster? Then you can build a two-degrees-of-freedom robot. Do you want it to sense its environment? Then you should add sensors to the front of the robot in the prototype area. Refer to the remaining chapters for ideas on how to expand your new creation. Remember, you can always buy more kits and make more robots!

References

[1] Technical Characteristics of Flexinol® Actuator Wires, Dynalloy, Inc.

[2] Gilbertson, R. G., *Working with Shape Memory Wires.* San Leandro, CA: Mondo-Tronics, Inc., 1992.

[3] Mills, J. W., "Area-Efficient Implication Circuits for Very Dense Lukasiewicz Logic Arrays," In *Proceedings of 22nd In-*

ternational Symposium on Multiple-Valued Logic, Sendai, Japan. IEEE Press: New York, 1992.

[4] Mills, J., and C. Daffinger, "An Analog VLSI Array Processor for Classical and Connectionist AI," In *Proceedings of Application Specific Array Processors,* Princeton, New Jersey. IEEE Press: New York, 1990.

[5] Brooks, R., "A robot that walks: Emergent behaviors from a carefully evolved network," *Neural Computation, 1*(2), 253–262, 1990.

[6] Beers, R., "An artificial insect," *American Scientist, 79* (September–October), 444–452, 1991.

[7] Wilson, E. O., *Sociobiology: The New Synthesis.* Harvard University Press, Cambridge, MA, 1975.

[8] Ballard, D. H., and C. M. Brown, *Computer Vision,* Prentice-Hall: Englewood Cliffs, NJ, 1982.

[9] Conrad, J. M., and J. W. Mills, *Stiquito: Advanced Experiments with a Simple and Inexpensive Robot,* IEEE Computer Society Press: Los Alamitos, CA, 1997.

[10] Conrad, J. M., and J. W. Mills, *Stiquito for Beginners: An Introduction to Robotics,* IEEE Computer Society Press: Los Alamitos, CA, 1999.

[11] Mims, F. M., *Engineer's Mini-Notebook: Schematic Symbols, Device Packages, Design and Testing,* Radio Shack, Fort Worth: TX, 1988.

Chapter 4 Appendix: The Printed Circuit Board and Soldering

The Stiquito Controller Board has been preprogrammed to make your Stiquito robot walk with one degree of freedom. Although this board does not include the materials to reprogram the microcontroller, several features were added to the board so that users could solder low-cost connectors to the board and be able to expand its capabilities. These features include:

- **Two-Degrees-of-Freedom Jumper.** If you shunt this jumper (J6), then the Stiquito Controller Board operates in two-degrees-of-freedom mode. If this jumper remains open, it runs in one-degree-of-freedom mode. You need to solder a jumper header connector to the board in position J6.
- **JTAG Connection.** The JTAG header (J2) is connected to a computer using a parallel port interface. To use the JTAG, you will need to solder a large header to the board at J2. You will also need to solder several small jumper headers to the board to support powering the Stiquito controller board. Using the JTAG is described in Chapter 5.
- **Using the JTAG Cable.** If the Stiquito Robot is attached to the Stiquito controller board, then to use the JTAG cable you must shunt jumper J5. You need to solder a jumper header connector to the board in position J5.
- **Powering the Stiquito Controller Board From the PC.** Putting a shunt on jumper J1 allows you to use the PC's power for the Stiquito controller board through the JTAG cable. Again, *do not* power the robot legs with PC power. You need to solder a jumper header connector to the board in position J1.
- **Reset Switch/Jumper.** When a software problem occurs, the reset line can be activated to reset the system to reinitialize the microcontroller. There is a jumper available that can support a switch or jumper

header. You need to solder a jumper header connector to the board in position J8.

The hardware for this additional functionality can be purchased from suppliers listed in the Appendix. If you implement any of this functionality, you must solder these connectors/jumpers to the board before you proceed.

Required Soldering Skills

Soldering mechanically and electrically connects components and integrated circuits to the interface card. The basic technique for soldering is described below [11] (see Figure 4.37). You may want to practice on an old circuit board before you solder the parts to the interface card.

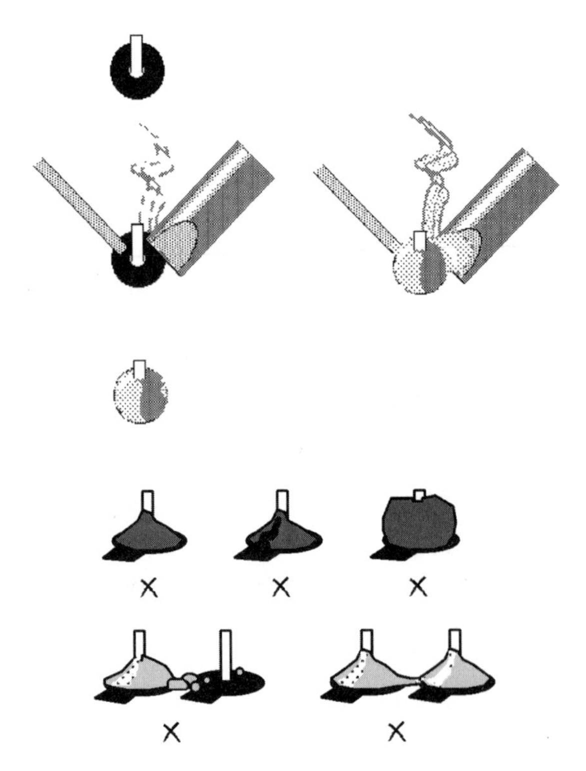

Figure 4.37 Soldering skills.

- Use the soldering iron's tip to heat the *pad,* not the integrated circuit pin. When the pad is hot, touch the solder to the heated pad, and the solder will flow onto the pad and the pin.
- Use just enough solder to wet the pin and cover the pad.
- Each solder joint should be bright, shiny, and have flowed evenly around the pin on the pad. The solder on adjacent pads must not touch.
- A solder joint should *not* be dull, cracked, or beaded up on the pad.
- A solder joint must *not* cross between two pads, or a pad and a trace. This will create a short circuit. Your interface card will almost certainly not work correctly.

After you have finished soldering, you should examine the board for workmanship errors (Figure 4.38). Check the board for short circuits and broken traces:

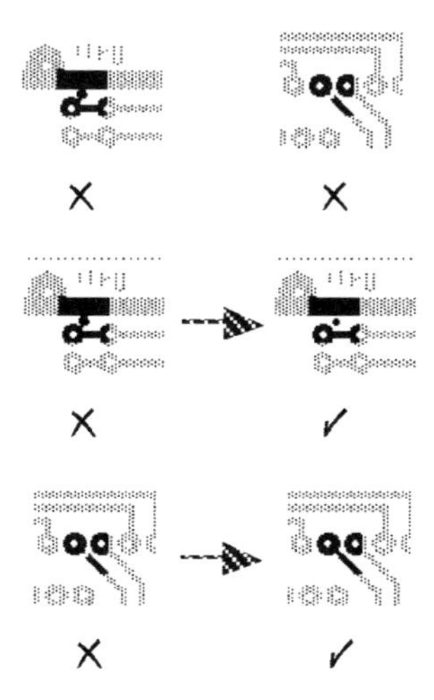

Figure 4.38 Examining the board for soldering faults.

- Examine the wiring side of the interface board. Look at places where one trace (wires on board) or pad (round circle on board) is near another; check that they do not touch. Look at long traces and near bends; check that the trace is not broken at that point.
- If traces or pads touch when they should not, use the knife to cut the unwanted connection.
- If a trace is broken, lightly sand it on either side of the trace, and then solder the broken ends together using a piece of fine wire to bridge the gap.

Precautions:

- Soldering irons get *very hot.*
- Do *not* touch exposed metal on a hot soldering iron. Hold it by the insulated handle.
- Do *not* lay a hot soldering iron on the work surface or flammable material.
- Use a soldering iron holder or lay a piece of wood under but not touching the tip to protect your work surface.
- Do *not* handle the integrated circuits by the pins.
- In winter and dry weather, touch a large metal object (table, door frame) to discharge any static electricity before handling integrated circuits.
- *Always* wear safety glasses when performing this work.

Placing and Soldering Connectors

If you only plan on running the Stiquito robot in two-degrees-of-freedom mode, you only need to solder one jumper connector at J6. If you plan on programming your Stiquito controller board, you should solder all connectors to the board. Figure 4.39 below shows the location of all jumper connectors.

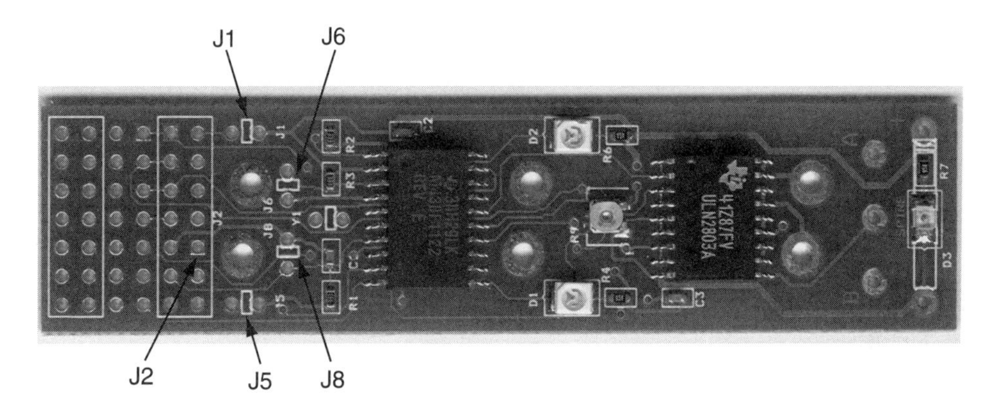

Figure 4.39 Location of jumpers on the Stiquito controller board.

The first step is to insert the four two-pin jumper connectors and the two-row, 14-pin connector into the board (Figures 4.40 and 4.41). Carefully turn the board over and solder all pins to the pads. Follow the soldering guidelines above to ensure you have not shorted two pads together. Ensure that each solder joint is shiny and completely surrounds each pin.

Figure 4.40 Placing the connectors on the Stiquito controller board.

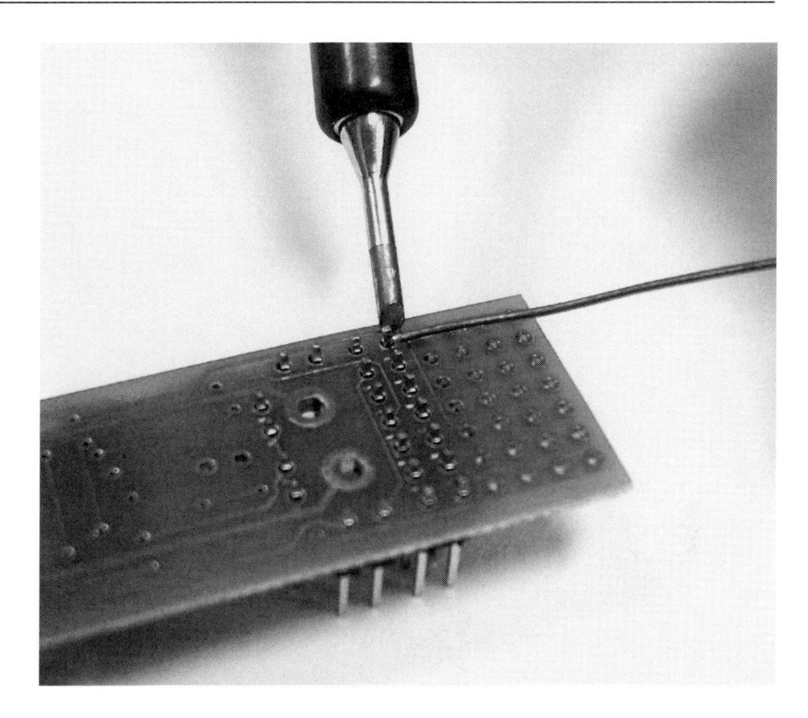

Figure 4.41 Soldering the connectors on the Stiquito controller board

Chapter 5

Stiquito Programming Using Texas Instruments MSP430F1122

The inclusion of a microcontroller board in this book makes this entire package a very valuable learning tool and environment. While designing the Stiquito microcontroller board, cost was the primary consideration, but features and expandability of the board were also important considerations. The microcontroller and other hardware components were selected based on the ability to drive a

ISBN 0-471-48882-8

walking Stiquito as well as the hardware's ability to be reprogrammed for other applications. The Stiquito microcontroller board has been preprogrammed with the following features and functionality:

- One-degree-of-freedom Stiquito—Movement is based on one-degrees-of-freedom movement of legs (two tripods), with LED flashes to indicate functionality.
- Two-degree-of-freedom Stiquito—Movement is based on two-degrees-of-freedom movement of legs (four tripods).
- Analog-to-digital (A/D) conversion—To control the speed of the Stiquito.
- Pulse-width modulation—For less power consumption during actuation of the legs.

This chapter explains how to program the Stiquito microcontroller board and discusses the features of the Texas Instruments (TI) MSP430F1122 microcontroller. It also explains the pin configuration of Stiquito and various features of the TI microcontroller.

Introduction

As explained in previous chapters, Stiquito is a self-sufficient, small, six-legged walking robot. It uses Flexinol® attached to the legs of the robot as its basis for movement.

The Stiquito microcontroller board (Figure 5.1) is designed for one-degree-of-freedom and two-degrees-of-freedom movement. One-degree of freedom means movement on the horizontal plane (back and forth) only, and two-degrees-of-freedom movement means movement on both the horizontal and vertical planes (both back and forth and up and down). Stiquito uses two sets of tripods for two-degrees-of-freedom operation. When one tripod is lifted off the surface, the other tripod is activated and moved back. The first tripod is then lowered, and the second is lifted and also returns to its forward position. Then the first is moved back. Finally, the tripod in the air is lowered. The process repeats and the Stiquito robot moves

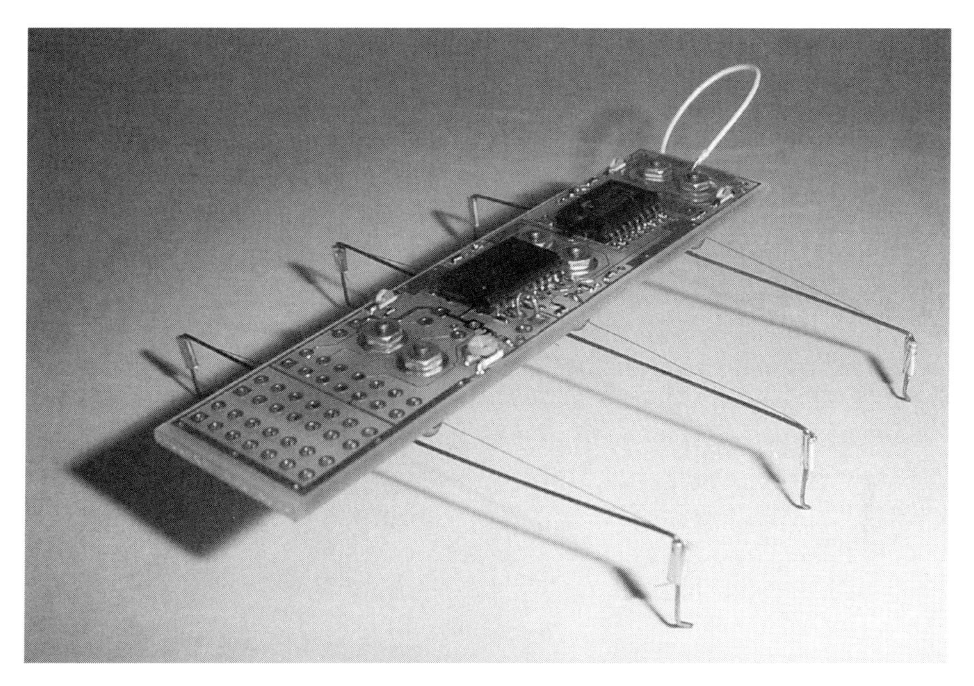

Figure 5.1 Stiquito robot and Stiquito microcontroller board.

forward without friction, which differs from one-degree-of-freedom movement.

The Stiquito printed circuit board is controlled by the TI MSP430F1122 programmable microcontroller. This microcontroller is a member of the TI family of ultra-low-power, mixed-signal microcontrollers. It has a built-in 16-bit timer and a 10-bit, 200-kbps A/D converter with internal reference, sample-and-hold, autoscan, data transfer controller, and fourteen I/O pins. The MSP430F1122 has 256 bytes of RAM and 4 Kbytes + 256 bytes of flash memory. The flash memory can be programmed via the JTAG port using the development kit or in-system by the CPU. The microcontroller selected for this board has a total of 20 pins and two I/O ports [1]. Port 1 has eight I/O pins and Port 2 has six I/O pins. For this board, the following pin assignment has been made:

Port 1

P1.0: Tripod B 2-dof control	output
P1.1: Tripod A 2-dof control	output
P1.2: Tripod B control	output
P1.3: Tripod A control	output
P1.4 through P1.7: JTAG	input

Port 2

P2.0: ADC input from POT to adjust speed	input
P2.0 through P2.2: routed to protype area	input/output
P2.3: Tripod B LED	output
P2.4: Tripod A LED	output
P2.5: 2-dof jumper	input

These assignments and the rest of the circuit are shown in the schematic in Chapter 2.

Specifications for the Operation of the Controller Board

Every engineering product needs to have operating specifications, and this board is no exception. General specifications for the board were presented in Chapter 2. Additional operational specifications for this controller board are:

1. Power consumption is very important; therefore, control registers should be set to reduce power consumption.
2. The slowest clock should be selected (i.e., BCSCTL1 and DCOCTL should have RSEL = 000 and DCO = 000, for a clock less than 100 kHz).
3. The board should decide to execute one-degree-of-freedom versus two-degrees-of-freedom operation within one second after the board is powered on or reset. If P2.5 = "1," run in one-degree-of-freedom

mode; if P2.5 = "0," run in two-degrees-of-freedom mode.

4. The board should use pulse-width modulation (PWM) so as to never hold a transistor pair high for more than 32 ms.
5. When running in two-degrees-of-freedom mode, it should only activate one transistor pair at a time.
6. Timers and interrupts should be used to determine the activation of the transistor pairs, not "wait loops."
7. Input from the potentiometer is checked at pin 8, which is a test input to the A/D converter of the TI MSP430F1122.
8. The period for one complete cycle (both tripods activated and relaxed) will range from three seconds up to 11 seconds in 0.25 second increments. Therefore, 32 different states should be possible.

Specification 1 is met by using the low-power-mode bits:

```
LPM0_bits      (CPUOFF)
```

Specification 2 is met by setting the proper speed registers:

```
BCSCTL1 &= ~(RSEL0 + RSEL1 + RSEL2); // Sets RSELx
   bits to 000, slowest speed
DCOCTL &= ~(DCO0 + DCO1 + DCO2); // Sets DCOx
   bits to 000, slowest speed
```

Specification 3 is met by checking pin P2.5 within one second after the processor starts executing code. If the pin is logic zero, the software will identify the board to run in two-degrees-of-freedom mode.

Specifications 4 and 5 are met by changing the states in a way that, at any time in the state diagram, no more than one pair of transistors will be held high. Power saving was done by using PWM. The waveform for output HIGH in the PWM mode is shown below in Figure 5.2.

Specification 6 is met by using Timer_A for the purpose of introducing delay. Timer_A is switched on whenever a

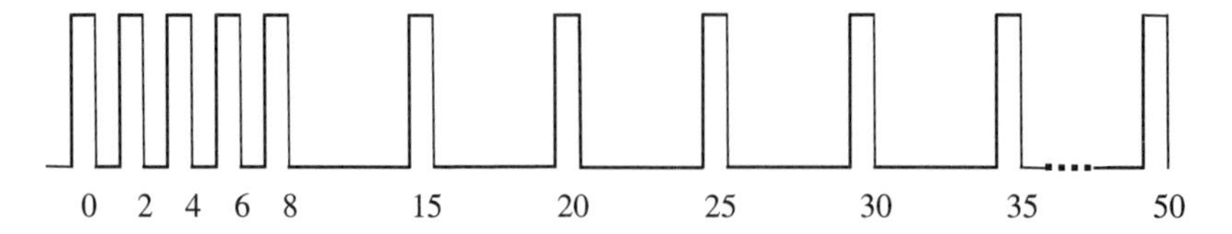

Figure 5.2 Waveform for output HIGH in the PWM mode.

delay period needs to be introduced, during which all operations are stopped. After generating the required amount of delay, the timer is switched off.

To meet specification 7 (speed control), input voltage to the 10-bit A/D converter of the MSP430F1122 was varied by changing the resistance using the potentiometer. Depending on the voltage level given to the A/D converter, the software decides the speed of Stiquito's movement.

Specification 8 is also achieved by carefully setting appropriate values of timers.

Software Implementation

Stiquito's microcontroller board code was written in C language. It was compiled using the free "IAR Embedded Workbench IDE 3.0B (3.0.2.0)," which is a part of tool "Kickstart V2" [2]. In Chapter 4, we showed how to attach the JTAG connector to the Stiquito controller board. To program the controller board, you will need to attach the JTAG cable (Figure 5.3) to the board connector, and then plug the cable into the parallel port connector of your PC. The tool "Kickstart V2" combines several tools including a C-language compiler, assembly language assembler, debugger, and a simulator. Techniques like inserting a break point, watch, and step-by-step execution can be used in the IDE environment to debug the program. The website in Reference 1 will provide instructions on how to load the IAR Embedded Workbench software on your PC, and will also provide a tutorial for using the software. There are also other excellent books on programming the MSP430 family of microcontrollers [3]. An example of the

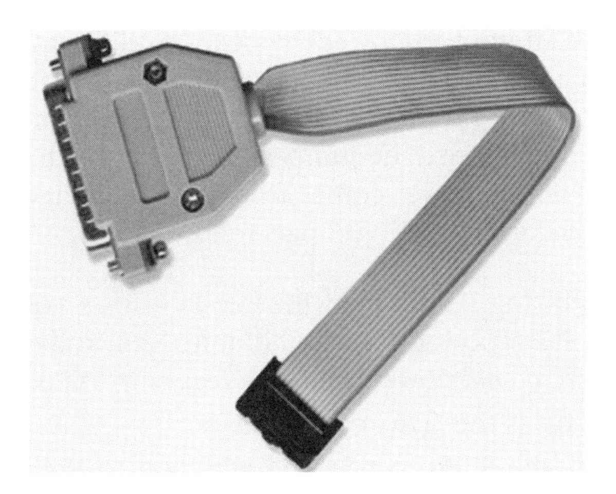

Figure 5.3 MSP430-JTAG for programming and flash emulation (courtesy of Olimex Ltd).

Stiquito controller board development setup is shown in Figure 5.4.

The Stiquito controller board can be powered from your PC through the JTAG cable, or it can be powered by attaching it to batteries. However, remember that you

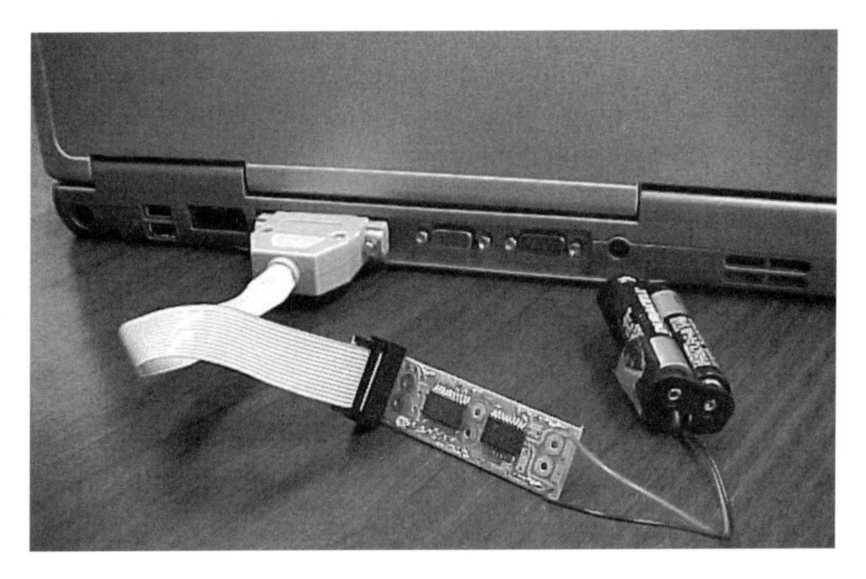

Figure 5.4 The Stiquito controller board development environment—attached to a PC through the PC parallel port.

should not attempt to power the board **and** Stiquito Flexinol® actuators. A PC cannot provide enough current to drive the legs, and you may damage your PC circuitry if you try. You can use the jumpers on the board to control where electrical power comes from and what devices use it. An explanation of all jumpers follows:

- J1: Putting a shunt on jumper J1 allows you to use the PC's power for the Stiquito controller board. Again, *do not* power the robot legs with PC power.
- J2: This is the JTAG connector
- J5: If the Stiquito robot is attached to the Stiquito controller board, then to use the JTAG cable you must shunt jumper J5.
- J6: If you shunt this jumper, then Stiquito operates in two-degrees-of-freedom mode. If this jumper remains open, it runs in one-degree-of-freedom mode.
- J8: This is a direct line to the RESET pin on the microcontroller. If you download new code and a software problem occurs, the reset line can be activated to reset the system to reinitialize the microcontroller. This jumper can support a switch or jumper header.

See Figure 5.5 for the location of these jumpers. See Chapter 4 for instructions on adding these jumper headers.

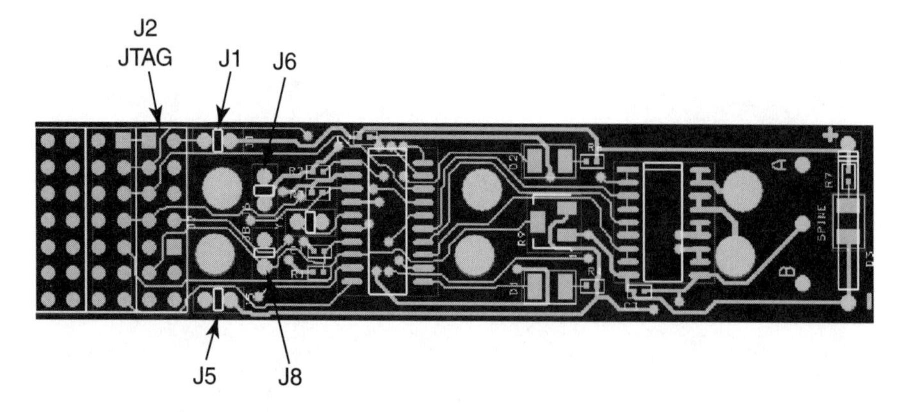

Figure 5.5 The Stiquito controller board layout showing the location of important jumpers.

Programming the MSP430

The main objectives of the Stiquito controller board software are to:

1. Ensure low power consumption during Stiquito operation
2. Provide the ability for users to operate Stiquito with one-degree-of-freedom or two-degrees-of-freedom movement
3. Provide a speed control mechanism for the user, which sets the speed of Stiquito movement
4. Meet all operating specifications for the board

Software Architecture

A software architecture forms the backbone for building a successful software system. An architecture largely permits or precludes a system's quality attributes such as performance or reliability. The right architecture is the determining factor for software success, whereas the wrong one is a recipe for disaster. After gathering all the requirements for Stiquito and finalizing the hardware components that were going to be used, a careful analysis resulted in developing the architecture shown in Figure 5.6.

The following components constitute the major functions of the Stiquito controller board software architecture.

Main Process: main.c. Initializes the input/output ports, watchdog timer, and interrupts:

- Initialize respective port pins to input/output mode
- Initialize watchdog timer and disable
- Enable global interrupts
- Enter low power mode with interrupt support

In order to initialize various registers, the following commands are used:

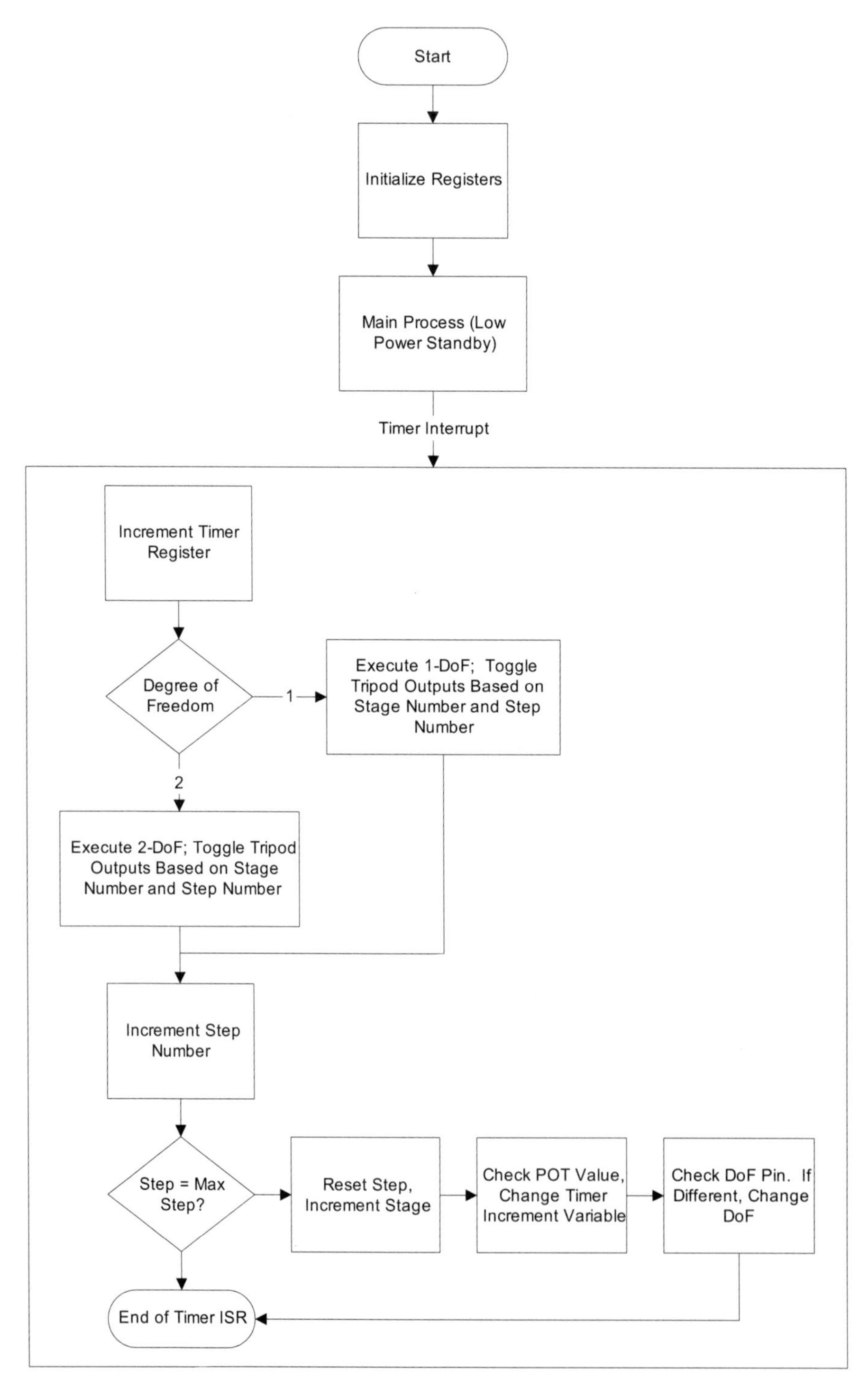

Figure 5.6 Block diagram of software architecture.

BCSCTL1 &= ~(RSEL0 + RSEL1 + RSEL2) (sets resistor select bits to 000 for slowest clock speed)

DCOCTL &= ~(DCO0 + DCO1 + DCO2) (sets digital crystal oscillator bits to 000, slowest speed)

WDTCTL = WDT_MDLY_32; (sets watchdog timer to 32 millisecond interval. Note: Even though the watchdog timer is not used, this register must be set or the WDT will reset the board at a fixed interval)

CCTL0 = CCIE; (sets timer A0 to use the capture and compare register 0)

TACTL = TASSEL_2 + MC_2; (sets timer A0 into continuous count mode, using SMCLK clock input)

P1DIR = 0x0F. 0x0F is a hex number whose binary equivalent is 0000 1111. This command sets Port 1 in following manner:

Port 1.0: Tripod B—two degrees of freedom, output

Port 1.1: Tripod A—two degrees of freedom, output

Port 1.2: Tripod B—output

Port 1.3: Tripod A—output

P1OUT = 0x00. 0x00 is a hex number whose binary equivalent is 0000 0000. Turn off legs and other outputs by setting Port 1 output low.

P2DIR = 0x1E. 0x0E is hex number whose binary equivalent is 0000 1110. This command sets Port 2 in following manner:

Port 2.0: ADC input from POT to adjust speed

Port 2.2 and P2.0: routed to prototype area

Port 2.3: Tripod B LED, output

Port 2.4: Tripod A LED, output

Port 2.5: Two-degrees-of-freedom jumpers, input

P2OUT = 0x00. 0x00 is a hex number whose binary equivalent is 0000 0000. Turn off legs and other outputs by setting Port 2 output low.

P2OUT |= 0x18;

0x18 is hex number whose binary equivalent is 0001 1000. Turns off LEDs (inverse logic) while leaving other ports off set_new_speed(); (custom function that reads initial state of potentiometer, calculates speed of timer A0, and sets CCR0 register)

set_dof(); (custom function that checks pin 2.5 to determine 1 or 2 degree of freedom movement)

Interrupt Service Routine: Timer_A3(void). The interrupt service routine is triggered every time the timer counts up to the number determined by the set_new_speed function.

- This interrupt routine is invoked twice per pulse, once to turn on and once to turn off each pin that is active during any given stage
- Increments the CCR0 register to induce the next interrupt in the given interval
- Between each stage, executes the set_dof and set_new_speed functions to update speed and degree of freedom, if the degree of freedom jumper or speed potentiometer have changed
- Depending on where the robot is in the walking cycle, toggles pins related to the active legs

#pragma vector=TIMERA0_VECTOR
__interrupt void Timer_A3 (void) (defines the execution outlined above)

Function to toggle output pins: void toggle(int); This function toggles the tripod represented by the integer number passed to it. (ex: toggle(3); will toggle the port corresponding to port 1.3, the A tripod pull-back leg) If the leg being moved corresponds to one of the pull-back legs, the LED port paired with that tripod is also toggled.

Function to calculate power of a number: int power(int, int); This function calculates the power of a number. For the purpose of this program, the number will always be 2 and the power will be between 0 and 7. This function is used by the toggle function to generate masks for toggling the tripod output pins.

Function to measure position of potentiometer: int get_speed(void); This function activates the analog to digital converter, waits for a reading to be aquired, reduces the measurement to an integer from 1 to 32, and returns the resulting number. This function is used by the set_new_speed function.

Function to set a new speed based on potentiometer: void set_new_speed(void); This function calls get_speed to get a new speed, calculates the needed timer register value, and sets that value. Because of the large difference in possible speeds, this function also calculates the number of pulses each stage should have. This is necessary to satisfy requirement #4, that no pin should be on for more than 32ms.

Function to determine and set degree of freedom: void set_dof(void); This function reads pin 2.5 and sets a global variable accordingly. This variable is then read by the timer interrupt routine and the legs are moved accordingly.

Dependencies: msp430x11x2.h Header file.

Algorithms for Two/One Degree of Freedom

Figures 5.7 and 5.8 show the algorithm for the two degrees of freedom and one degree of freedom, respectively. In these algorithms, the potentiometer (Port 2.0 input) value is read between each state to decide the speed of the Stiquito robot. Timers and interrupts, not wait loops, determine the activation of the transistor pairs, which meets Specification 6. Tables 5.1 and 5.2 show the specific output port pin activations for two degrees of freedom and one degree of freedom, respectively.

Specification 7, speed control of the robot, was implemented using A/D conversion. The module implements a 10-bit SAR core, sample select control, reference generator, and data transfer controller (DTC). The ADC10 module made software implementation of Stiquito speed control

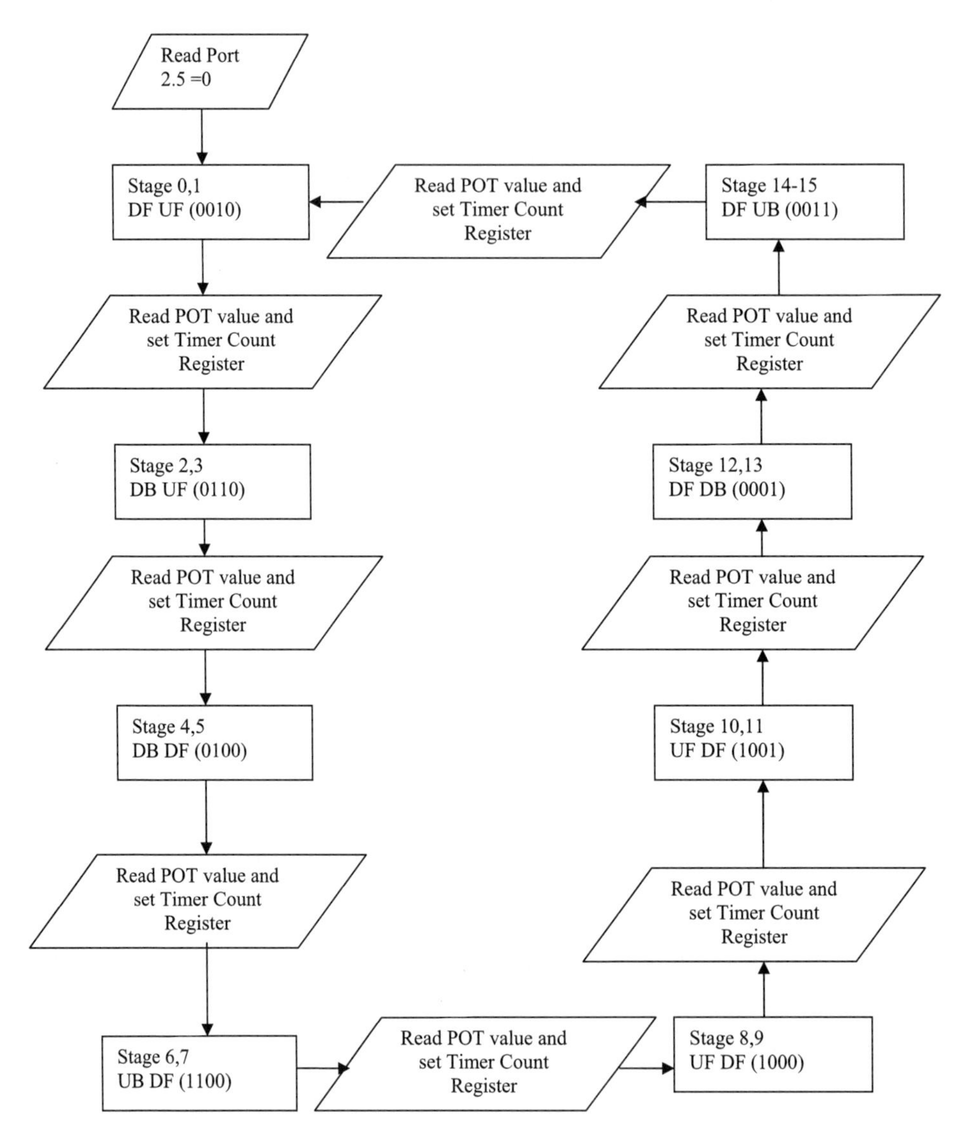

Note: Each state on this diagram includes multiple stages. The first stage in which a tripod is first turned on, it is pulsed at twice the regular speed. This quick pulsing lasts for one half of a state, therefore, even though the same legs are being activated, there are two stages in each state.

Figure 5.7 State diagram of two-degrees-of-freedom movement.

Table 5.1 Shows the status of the tripod A and tripod B pins in the different states for two degrees of freedom

	Two degrees of freedom		Tripod LED	
State	Tripod status	Pin	Active LED	Pin
1 0010 **DF UF**	Tripod B(up)	P1.0	Tripod A	P2.4
0000 **DF DF**	Rest State		Both Tripod LED off	
2 0110 **DB UF**	Tripod B(up) Tripod A(back)	P1.0 P1.3	Both Tripod LED on intermittently	P2.4 P2.3
0000 **DF DF**	Rest State		Both Tripod LED off	
3 0100 **DB DF**	Tripod A(back)	P1.3	Tripod B	P2.3
0000 **DF DF**	Rest State		Both Tripod LED off	
4 1100 **UB DF**	Tripod A(back) Tripod A(up)	P1.3 P1.1	Tripod B	P2.3
0000 **DF DF**	Rest State		Both Tripod LED off	
5 1000 **UF DF**	Tripod A(up)	P1.1	Tripod B	P2.3
0000 **DF DF**	Rest State		Both Tripod LED off	
6 1001 **UF DB**	Tripod A(up) Tripod B(back)	P1.1 P1.2	Both Tripod P2.4 LED on intermittently	P2.3
0000 **DF DF**	Rest State		Both Tripod LED off	
7 0001 **DF DB**	Tripod B(back)	P1.2	Tripod A	P2.4
0000 **DF DF**	Rest State		Both Tripod LED off	
8 0011 **DF UB**	Tripod B(back) Tripod B(up)	P1.2 P1.0	Tripod A	P2.4

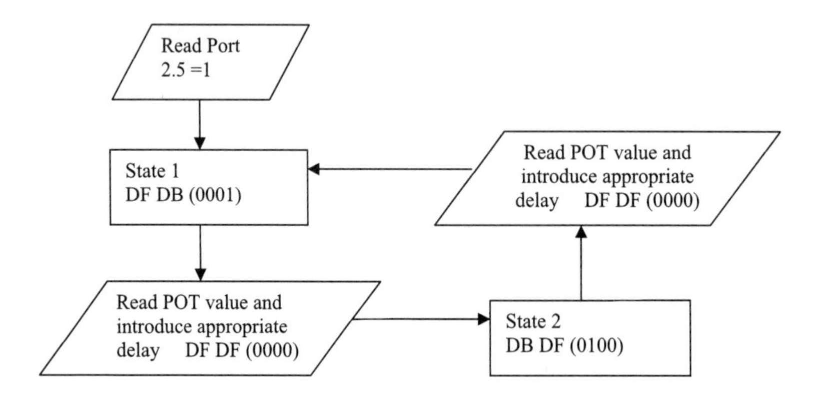

Note: Even though there are only 4 needed stages (2 for each tripod; 1 quick pulsing, 1 regular), all 16 are included to allow smoother transitions between 1 and 2 degree of freedom.

Figure 5.8 State diagram of one-degree-of-freedom movement.

Table 5.2 Shows the status of the tripod A and tripod B pins in different states for one degree of freedom

	One-degrees of freedom		Tripod LED	
State	Tripod status	Pin	Active LED	Pin
1 0001 **DF DB**	Tripod A (back)	P1.3	Tripod B	P2.3
0000	Rest state		Both Tripod LED off	
2 0100 **DB DF**	Tripod B (back)	P1.2	Tripod A	P2.4
0000	Rest state		Both Tripod LED off	

simple and was achieved in a few lines of code. During software implementation, a single sample was made on A0 with reference to Avcc. Software sets ADC10SC to start sample and conversion; ADC10SC is automatically cleared at end of conversion (EOC). The ADC10 internal oscillator times the sample (16 ×) and conversion. The ADC10BUSY flag is polled for EOC. The A/D conversion value is stored in ADC10MEM. Refer to *MSP-FET430 User's Guide* [4], Chapter 18, for details related to ADC10 module programming.

The code listing for the Stiquito controller board can be found at the IEEE Computer Society Press Stiquito support page, http://computer.org/books/stiquito.

Further Opportunities

The Stiquito controller board has been designed to make the Stiquito robot walk with two different gaits at various speeds. Since the microcontroller has flash memory, the existing program can be changed. Perhaps you want to experiment with different gaits. You can change the code and download the new program into the flash memory. Maybe you want to add a sensor to the prototype area of the board. You can solder the part to the board and then change the program to use the part. We have provided a wonderful environment for learning about robotics and embedded systems. Your use of it is only limited by your imagination!

References

[1] http://focus.ti.com/lit/ds/slas361c/slas361c.pdf (MSP430F1-122 Data sheet).

[2] http://focus.ti.com/docs/tool/toolfolder.jhtml?partnumber=msp430freetools.

[3] Nagy, C., *Embedded Systems Design using TI MSP430 Series,* Elsevier, 2002.

[4] *Texas Instruments MSP-FET430 Flash Emulation Tool (FET) User's Guide,* "MSP-FET430 Users Guide.pdf." Literature on MSP430 CD ROM—February 2002.

Resources

Stiquito controller board code listings: http://computer.org/books/stiquito

TI development tools: http://focus.ti.com/docs/tool/toolfolder.jhtml?partnumber=msp430freetools

Chapter 6

A Two-Degrees-of-Freedom Stiquito Robot

Stiquito Controlled! Making a Truly Autonomous Robot. By James M. Conrad
ISBN 0-471-48882-8

In this chapter, we discuss an extension to the classical Stiquito design to make it walk with steps rather than pushing itself along. In order to make Stiquito step, the user will need to turn it into a two-degrees-of-freedom robot. Due to the complexity of locomotion, this robot must include computer control. This design uses the popular BASIC Stamp 2 microcontroller* to coordinate Stiquito's twelve Flexinol® muscles.

Motivation for a Stiquito Variation

In the summer of 2001, a small engineering class at North Carolina State University was given the assignment to design and implement a functional robot for a robotic race. This race had specific design rules and restrictions that the students had to follow. The rules for the race and the requirements for the robot were that the robot:

- Must use Flexinol® for locomotion
- Must use legs in its propulsion (two or more legs required)
- Must walk four times its length in the fastest amount of time on a smooth Formica® surface
- Must have an on-board microprocessor
- Must stop walking via sensor input
- Must measure no greater than 12 inches long by 12 inches wide
- Would be given multiple trials, with the best time taken for competition comparison
- Would be allowed to have either external (tethered) or attached power supply

The summer course, "Simple Robots and Microprocessors," had concentrated on the Stiquito robot and the BASIC Stamp 2 microcontroller. Naturally, most students used Stiquito as the basis of their designs.

As described in Chapter 4, Flexinol® contracts when

*BASIC Stamp is a trademark of Parallax, Inc.

current runs through it, and then returns to its normal position when the current is removed and the music wire pulls the Flexinol® back into its normal position. This action causes a forward and backward pulling movement, hence the primary movements of the original Stiquito.

A two-degrees-of-freedom Stiquito uses the regular Stiquito design [1], with the addition of six more Flexinol® actuator wires that lift the legs off the ground. Two-degrees-of-freedom movement means movement on both a horizontal plane and vertical plane (both back and forth and up and down). Using the two tripods of Stiquito legs, tripod A is lifted off the ground while the leg is in the forward (relaxed) position. Then, tripod B is moved backward while on the ground. Next, tripod B is moved up into the air and forward, and tripod A is moved down onto the ground and backward. The previous steps are repeated, beginning with tripod A. The term "forward" means that the leg relaxes and returns to standard position. The term "backward" means that the Flexinol® tightens. Two-degrees-of-freedom robots have a lot less friction than one-degree-of-freedom robots (which make movements on the horizontal plane only), and, therefore, move faster.

Course students decided that a microprocessor-controlled, two-degrees-of-freedom Stiquito (using screws) was the most time-efficient and cost-effective racebot. Designing the two-degrees-of-freedom Stiquito or any robot requires three parts:

1. The first and major part is the mechanics, which consists of the whole frame of the robot. This includes how the robot will look and function. The mechanics are very important, since this is the basis for the other two design parts. If the mechanics do not work, the robot does not work.
2. The second part is the hardware. The hardware consists of the microprocessors, chips, and circuitry needed to control the robot.
3. The third and last part of the design is the software. The software is not required if the robot is built with analog circuitry. However, software is needed for ro-

bots using a microprocessor. The software controls the microprocessor, which in turn drives the robot. Making the microprocessor software control the robot efficiently (i.e., consume as little current as possible) was an important goal.

Mechanical Design

The frame for the two-degrees-of-freedom Stiquito racebot uses the original one-degree-of-freedom Stiquito frame. The robot has six legs, uses a tripod gait, but also uses two degrees of freedom. A tripod gait is the movement of three legs at the same time—the front and back legs on one side and the middle leg on the other side. Refer to Chapter 4 for the one-degree-of-freedom Stiquito design. For a two-degrees-of-freedom Stiquito, an extra 55 mm of Flexinol® is attached to each leg. The extra wire is screwed to the upper surface of the Stiquito body for vertical movement. See Figure 6.1 for a close-up view of these connections. Two extra 28-AWG wires are attached to the bottom of the body to control the vertical tripod gait movements. A small piece of rubber is attached to the end of each foot to allow it to grip onto the surface. Two holes are drilled at

Figure 6.1 A two-degrees-of-freedom Stiquito.

each end of the body to attach the printed circuit board on top of the Stiquito.

Electrical Design

The printed circuit board is a regular epoxy board ("perf board") with several one-millimeter holes used to construct circuitry from scratch. The circuit board contained a microcontroller, a ULN2803A chip, a bump/switch sensor, and sockets, all connected according the circuit schematic.

The BASIC Stamp 2 microcontroller (see Figure 6.2), sold by Parallax Incorporated, was used to control the legs of Stiquito. The BASIC Stamp 2's dimensions are $1\frac{3}{16}$ inches by $\frac{5}{8}$ inches by $\frac{3}{8}$ inches, and weighs only 0.02 pounds. The BASIC Stamp 2 contains a PIC16C57 (interpreter chip) microcontroller, runs at the speed of 20 MHz, uses 2 kilobytes of EEPROM, and includes 32 bytes of RAM [2, 3]. The microcontroller interprets BASIC language programs for instructions. The BASIC Stamp 2 has

Figure 6.2 The BASIC Stamp 2 microcontroller contains a complete microprocessor system, including RAM, EEPROM, an operating system, a serial port, and 16 input/output ports.

16 input/output ports and a total of 24 pins. It needs 7 mA of current to run, and 50 μA of current to sleep. It can also provide source current of 20 mA and sink 25 mA of current.

The ULN2803A, eight-transistor Darlington array chip [4], was used as a source of current from a battery to each tripod since the current from the BASIC Stamp is not enough to contract the Flexinol®. The ULN2803A chip has a total of 18 pins. Pin numbers 9 and 10 are used for ground and power, respectively. Pins 1–8 are used for inputs, and pins 11–16 are used for outputs. Each input pin corresponds to the output pin directly across from each input pin. Refer to Figure 6.3 for pin numbers. The chip input has an internal 2.7 kohm resistor and requires at least 5 volts to run.

The rules of the race required that each robot had a

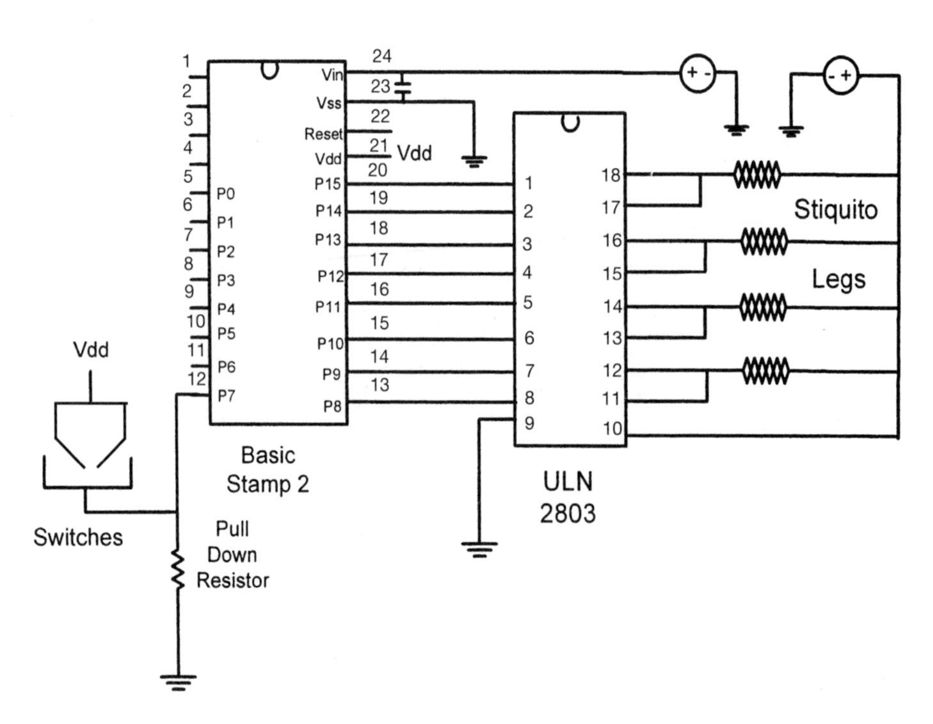

Figure 6.3 A schematic of the Stiquito controller circuit shows how simple it is. There are only two components: The BASIC Stamp 2 and a ULN2903A Darlington transistor array.

mounted sensor to stop the robot from walking. The sensor used for the two-degrees-of-freedom Stiquito robot was a bump/switch sensor. The sensor was connected in parallel to a starting switch. The robot started when the start switch (DIP switch) was pressed, and it stopped when the sensor switch was pressed.

Four sockets were soldered onto the printed circuit board. The purpose of the sockets was to easily remove the BASIC Stamp 2 and ULN2803A chip for testing and debugging. One of the sockets had the same dimensions and pins as the BASIC Stamp 2, and the other socket had the same dimensions and pins as the ULN2803A chip. The other two sockets (two pins) were power sockets used to attach the battery to the board. The power used was two 9 volt batteries. One battery drove the BASIC Stamp and ULN2803A chip, and the other battery drove the legs. The batteries were attached to the printed circuit board by way of four 34-AWG magnet wires. A schematic of the electrical design is shown in Figure 6.3.

Software Design

The program was written in Parallax's BASIC programming language. The software was used to control the movements of the legs, to control the frequency of the transmission of current to the legs, and to detect sensor inputs. A BASIC Stamp is a single-board computer that runs the Parallax PBASIC language interpreter in its microcontroller. The developer's code is stored in an EEPROM, which can also be used for data storage. The PBASIC language has easy-to-use commands for basic I/O, like turning devices on or off, interfacing with sensors, and so on. More advanced commands let the BASIC Stamp interface with other integrated circuits, communicate with each other, and operate in networks. Table 6.1 shows examples of some of the Basic Language instructions available for use on the BASIC Stamp 2.

Flexinol® requires quite a bit of current. Running Flexinol® robots can drain a 9 volt battery in 15 minutes. It also needs cool ambient air to return to its normal po-

Table 6.1 A few of the BASIC Stamp instructions

`for x = n1 to n2`	Typical for loop, with a next at the end (i.e., `for blink = 1 to 100`)
`goto label`	Jump to `label` (i.e., `goto reblink`, label would be `reblink:`)
`if x=n then label`	Based on true equation, got to label (i.e., `if n=1 then reblink`)
`input` **X**	Configure port X as an input port (i.e., input 1)
`in`**X**	Read port x (i.e., `blink = in3`)
`next`	Ends for loop
`output` **X**	Configure port X as an output port (i.e., output 2)
`out`**X**	Make port X output a value (as in out0 = 0)
`pause xx`	Pause for xx milliseconds (`pause 1000` is pause for 1 second)
`x var word`	Create a variable called x, for use later (i.e., `blink var word`)

sition quickly. To compensate for these drawbacks, the software uses pulse-width modulation. By using pulse-width modulation, the software is letting just enough current to go through per time unit to contract the Flexinol®. Pulse-width modulation is performed by sending a pulse to the Flexinol® for a few milliseconds, pausing for a few milliseconds, and repeating the pulse again. In order to quickly heat (and contract) the Flexinol®, a pulse of 20 milliseconds on and 20 milliseconds off is required. This is done five times. To sustain the Flexinol® contraction, a pulse of 20 milliseconds on and 80 milliseconds off is required. This is done eight times. In order to move a tripod, a jumpstart pulse and a sustaining pulse are executed. See Figures 6.4 and 6.5 for details on how we implemented pulse-width modulation for this robot.

Many embedded systems can be implemented by using a state machine to identify the operating environment and how inputs and internal events (like timers) cause

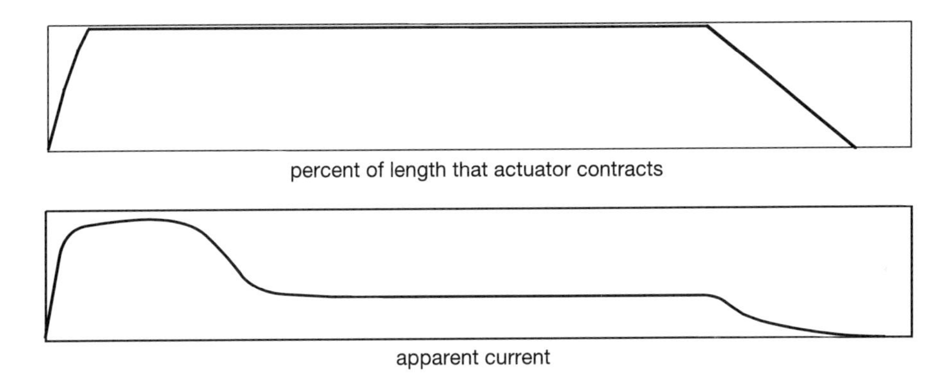

Figure 6.4 The Flexinol® actuator responds to the heat generated by the current pulses, and the loss of heat due to convection of the wire.

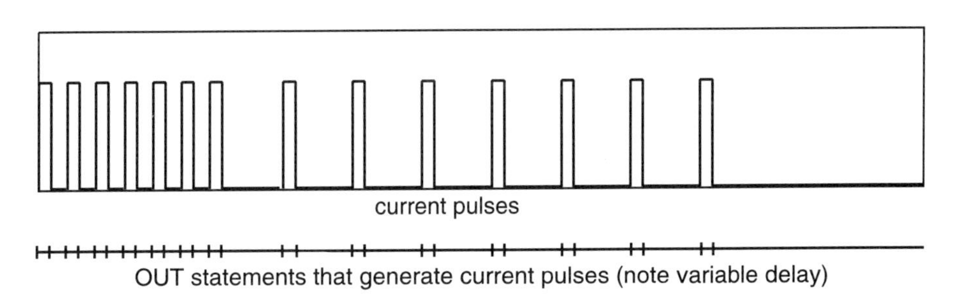

Figure 6.5 To save battery power and protect the Flexinol® wire, we use pulse-frequency modulate the current going to the Flexinol® wire.

transitions between states. We use a state diagram to show the transition between these states and identify the control of the legs in each state. Since there are four movements per tripod, there are four states per tripod. These are shown in Table 6.2. A "1" means the Flexinol® is activated. The state machine requires one initial state (all Flexinol® relaxed) and eight operational states. The different states are represented visually in Figure 6.6. Follow the state diagram for programming. Refer to the end of this chapter for the source code.

The software is looping in the beginning of the code, waiting for the "start switch" to be pressed (Port 7). After

Table 6.2 A description of the entries in the Stiquito state diagram

Designation			Description
Down and forward	DF	00	This is the relaxed state for all legs.
Down and back	DB	01	The legs are on the ground and pushing the robot forward.
Up and back	UB	11	Lift the legs off the ground, so when they go forward they do not pull the robot back
Up and forward	UF	10	With the legs out of the way, return them forward to prepare to place them on the ground

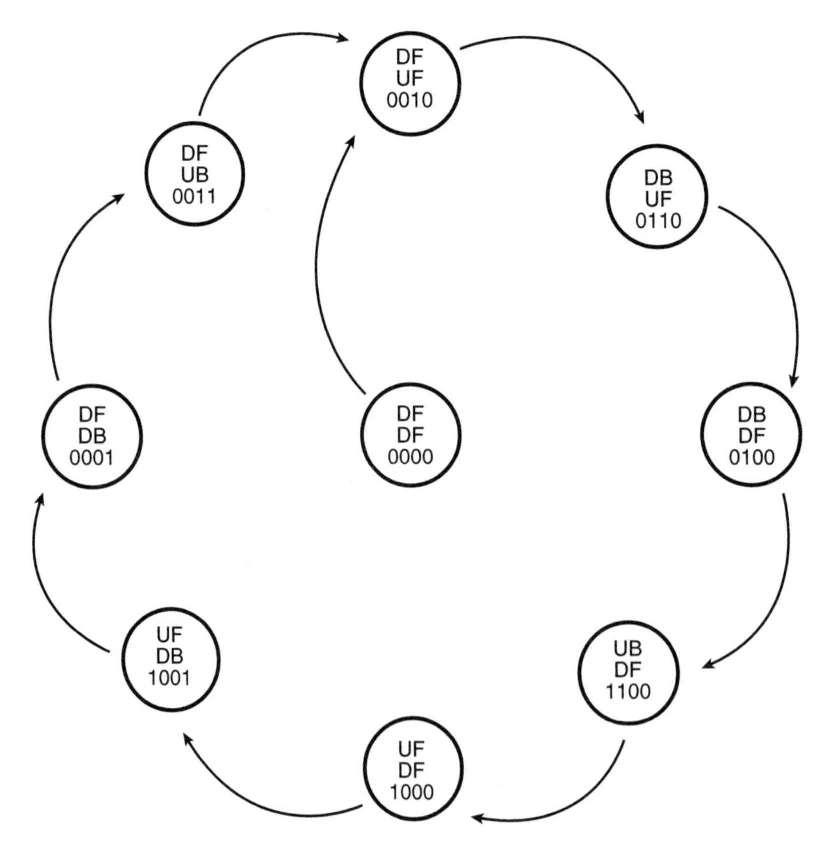

Figure 6.6 The Stiquito state diagram. The top letters represent one tripod; the middle letters represent the other tripod. The four binary digits represent which of the four groups of Flexinol® wires are activated.

the port registers a digital "one," the leg algorithm starts executing. In order to detect the bump sensor more frequently, the software polls Port 7 every pulse. After a digital "one" has been detected by the bump sensor, the program stops the leg algorithm and loops waiting for a digital "one" in Port 7 to start the legs again.

Implementation

In addition to the materials provided in the original Stiquito kit, the following parts were used. Most materials can be obtained from Micro Robotics Supply. See the Appendix for supplier contact information.

BASIC Stamp 2 microcontroller and socket (from Parallax, Inc.)
Brass hardware: 12 screws, 2 long screws, 14 washers, 30 nuts
20-AWG copper wire for sensor
Extra Flexinol® for the second degree of motion
1 printed circuit board (Stiquito's length by approximately twice Stiquito's width)
ULN2803A chip and socket
0.1 μF capacitor
Two two-pin sockets for battery
LEDs for testing
Pull-down resistor (1 kohm)
DIP switch
Soldering tools, wiring tools
Two 9 volt batteries

We followed the following steps to build the electronics of the robot:

1. Solder pins 13 through 20 (Ports 8 to 15) of the BASIC Stamp 2 socket to pins 8 through 1 of the ULN2803A socket, respectively. Solder every two pins together on the ULN2803A socket, starting with

pin 1. This connects two transistors together for more current to the legs. The result from the ULN2803A is four outputs for the two tripods with two-degree movements.
2. Pin 10 of the ULN2803A socket is soldered to the power of the battery number two socket. The VIN pin of the BASIC Stamp 2 socket is soldered to the power of the battery number one socket.
3. Pin 9 of the ULN2803A socket, VSS pin of the BASIC Stamp 2 socket, the ground of battery number two socket, and the pull-down resistor are soldered to the ground of battery number one socket.
4. The 34-AWG wire and one pin of the DIP switch (two pins) are soldered to the VDD of the BASIC Stamp 2 socket.
5. Solder a 0.1 μF capacitor between pins 23 and 24 of the BASIC Stamp.
6. The 20-AWG, the other pin of the DIP switch, and the power of the pull-down resistor are connected to pin 12 (Port 7) of the BASIC Stamp 2. The pull-down resistor pulls the pin's voltage down to its default voltage, since the pin's voltage tends to vary in an unknown voltage.
7. After all connections have been soldered, pin 12 (Port 7) of the BASIC Stamp 2 will be used as a detection pin. Every two pins starting with pin 13 through pin 20 will control one movement and one tripod.

It took roughly six hours to build the two-degrees-of-freedom Stiquito mechanics. The board took another six hours to produce. Testing and optimizing the Stiquito took yet another six hours. The software took 15 minutes to code. Total estimated time was 18 hours. The completed robot is shown in Figure 6.7.

While building this robot, we learned several tips that future builders should remember as well:

- The key to making the Flexinol® contract well (pulling 7–8 mm of music wire backward) is sanding

Figure 6.7 The completed two-degrees-of-freedom Stiquito with the microcontroller attached was able to walk 30 centimeters in one minute, compared to about 10 centimeters a minute for a one-degree-of-freedom Stiquito. The power tether is not shown in this photo.

the Flexinol® well. When sanded well, the Flexinol® will appear white and make a better connection.

- It is very important to test your code on a breadboard before you put it on your Stiquito. The breadboard should consist of BASIC Stamp 2, ULN2803A chip, and LEDs acting as legs. Not doing so may damage the Stiquito's Flexinol®.
- Do not tighten the vertical Flexinol® actuators too much. If this is done, the horizontal Flexinol® actuators break very easily.
- Check your board with a voltmeter. Be careful not to short your circuit connections on the board.
- Do not invert power and ground inputs!
- Test Flexinol® actuators with 3 volts; 9 volts will burn a single Flexinol® actuator.

- The 28-AWG wire and 34-AWG mag wires will snap easily; be careful when twisting or turning them at a joint.
- Flexinol® can get hot when current is running through it. Be careful not to burn yourself.

Racing Results and Future Plans

This two-degrees-of-freedom Stiquito won second place in the race; it was able to walk 30 cm in 61 seconds. The first-place robot won with an elapsed time of 41 seconds (and is the subject of a Chapter 7). The third place robot took 61.6 seconds to walk the 30 cm. The reason for the second-place finish was due to implementation of the state diagram in software. The Flexinol® contracted too long for the sustaining pulse. The current should only have been pulsed five times and not eight times. Also, not enough time was given for the Flexinol® to cool in certain states. The temperature in the racing room was very hot, and the Flexinol® could not relax fast enough.

Future plans are to improve the software for the robot, especially to achieve a better implementation of the state diagram. We also noticed a direct correlation between the ambient temperature and Flexinol® relaxation time, so adding a simple thermal sensor to distinguish between 20°C and 30°C could further optimize the gait.

References

[1] Conrad, J. M., and J. W. Mills, *Stiquito for Beginners: An Introduction to Robotics,* IEEE Computer Society Press: Los Alamitos, CA, 1997.

[2] Parallax, Inc., *BASIC Stamp Manual,* Version 2.0, Parallax, Inc., Rocklin, CA, 2004.

[3] BASIC Stamp 2 Microcontroller Information: http://www.parallax.com.

[4] ULN2803A Chip Information: www.st.com/stonline/books/ascii/docs/1536.htm.

Sample Software

This software shows the BASIC Stamp setup and the states of the walking Stiquito. It was written by Jon Williams of Parallax, Inc., based on the original code written by Scott Vu. This code and other examples can be viewed online at the IEEE Computer Society Press website, http://computer.org/books/Stiquito. Look for the *Stiquito Controlled!* selection.

```
' ==========================================================================
'   File...... Stiquito_JW.BS2
'   Purpose... Stiquito Leg Controller
'   Author.... Jon Williams - Parallax, Inc.
'   E-mail....
'   Started...
'   Updated... 11 JUL 2004
'
'   {$STAMP BS2}
'   {$PBASIC 2.5}
'
' ==========================================================================

' —[ Program Description ]——————————————————-

' —[ Revision History ]——————————————————

' —[ I/O Definitions ]———————————————————-

Legs            VAR     OUTH            ' leg control pins (P8 - P15)
Sensor          PIN     7               ' bump sensor input

' —[ Constants ]———————————————————————-
IsLow           CON     0
IsHigh          CON     1

No              CON     0
Yes             CON     1

N               CON     0
Y               CON     1

JmpTable        CON     %10111011       | states that require Jump_Start

' —[ Variables ]———————————————————————-
state           VAR     Nib             ' current state of walk cycle
idx             VAR     Nib             ' loop control
legsOut         VAR     Byte            ' leg outputs for current state
doJump          VAR     Bit             ' test bit for jump start
```

```
' —[ EEPROM Data ]——————————————————————————-
                                              ' Codes for Leg movements
                                              ' ————————————-
                                              ' D = Down
                                              ' U = Up
                                              ' F = Forward
                                              ' B = Backward

' Leg Outputs              111111             ' Tripod 1 / Tripod 2
'                          54321098
'                          ||||||||
S0010          DATA       %00001100           ' DF / UF
S0110          DATA       %11001100           ' DB / UF
S0100          DATA       %00000000           ' DB / DF
S1100          DATA       %00000000           ' UB / DF
S1000          DATA       %00000000           ' UF / DF
S1001          DATA       %00000000           ' UF / DB
S0001          DATA       %00000000           ' DF / DB
S0011          DATA       %00000000           ' DF / UB

' —[ Initialization ]————————————————————
Reset:
  Legs = %00000000                            ' legs off to start
  DIRH = %11111111                            ' make leg pins outputs

' —[ Program Code ]——————————————————————
Main:
  DO : LOOP UNTIL (Sensor = IsHigh)           ' wait for sensor
  PAUSE 500                                   ' for hardware synchronization
  state = 0

Walk_Cycle:
  DO
    READ (S0010 + state), legsOut
    LOOKUP state, [Y,Y,N,Y,Y,Y,N,Y], doJump
    IF (doJump = Yes) THEN
      GOSUB Jump_Start                        ' jump start if needed this state
    ENDIF
    GOSUB Move_Legs                           ' run normal movement code
    state = state + 1 // 8                    ' point to next state
  LOOP WHILE (Sensor = IsLow)                 ' run until sensor input

ReSync:
  PAUSE 5000
  GOTO Main

  END


' —[ Subroutines ]———————————————————————-
' Note: Only use jump start pulse to bring legs up

Jump_Start:
  FOR idx = 1 TO 5
    Legs = legsOut
    PAUSE 20
    Legs = %00000000
    PAUSE 20
```

```
    IF (Sensor = IsHigh) THEN EXIT
  NEXT
  RETURN


' Standard leg movement code - energy saving
Move_Legs:
  FOR idx = 1 TO 8
    Legs = legsOut
    PAUSE 20
    Legs = %00000000
    PAUSE 80
    IF (Sensor = IsHigh) THEN EXIT
  NEXT
  RETURN
```

Chapter 7

Optimizing the Stiquito Robot for Speed

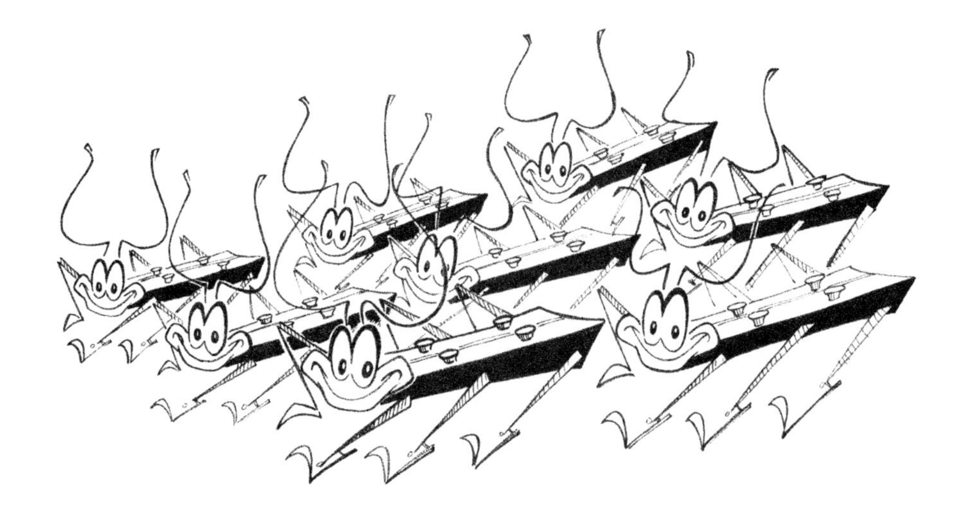

As part of North Carolina State University's Introduction to Robotics course, the enrolled students designed, built, and raced Flexinol®-based robots. One group started out building a basic Stiquito robot (Figure 7.1), and then looked at many factors affecting the speed. Some of the factors they examined were leg length, number of legs, degrees of freedom, gait motion, and the software used to animate the legs.

ISBN 0-471-48882-8

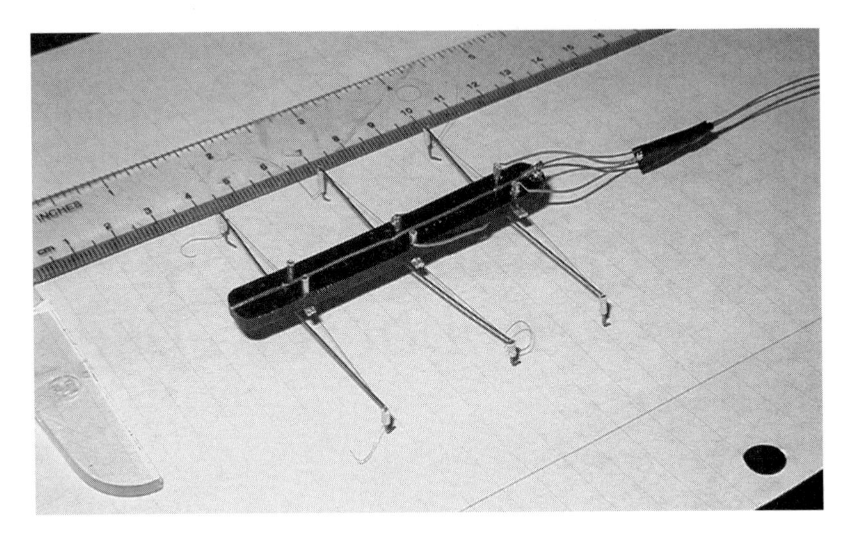

Figure 7.1 The classic Stiquito robot, built with crimps.

The Stiquito robot uses about 40 mm of Flexinol® wire to actuate each of its legs. The directions in this book and previous Stiquito books [1, 2] have the builder create a robot with a hard-wired tripod gate. The kit in these books comes with a manual controller, but additional kits can be bought from the publisher to control Stiquito via a computer's parallel port [1], or with a Parallax BASIC Stamp 2 processor, as shown in Chapter 6.

The students built a total of five robots over the course of the semester. They experimented with the code and finally decided that a robot having six long legs with two degrees of freedom was the best. They determined that the gait with the best speed was the one that used the Flexinol®-contraction phase to push the robot forward. Their experimentation with different designs paid off—their final robot won the race!

Motivation and Observations

After building their first Stiquito robot with the standard kit, the class suspected that there must be many ways to optimize Stiquito's speed. They observed that there are several factors that could easily influence the speed: the friction of the legs as they move, the number of legs that move, whether the robot moves forward during the contraction or relaxation of the Flexinol®, the lengths of the legs, and the speed of the leg contractions. However, the race rules constrained the length of the robot: the race distance was four times the length of the robot.

In the one-degree-of-freedom robot, the legs drag on the ground as they move forward. It is highly likely that this friction slows the robot, so the first modification attempted was legs with two degrees of freedom. The gait of a classical Stiquito follows the following motion:

1. The Flexinol® contracts and the body moves forward, dragging the noncontracting legs along.
2. The contracted legs then relax and scrape the surface. The noncontracting legs with the pointed feet grip the surface and keep the body from moving back.
3. The other set of legs contract, dragging the first set along, and the whole procedure repeats.

If the robot is built with two degrees of freedom (like in Chapter 6), instead of dragging forward, the legs would lift off the ground during the relaxation (see Figure 7.2).

Another friction-related effect that was noticed is that as the robot propels itself forward, the feet slip backward on the ground. Ratcheting the feet as recommended in the previous Stiquito books [1, 2] helped this somewhat, but the feet still skidded. Students tried using ThinkPad mouse tips for the feet, but they were too sticky and the robot didn't move at all. They ended up putting small dots of hot glue at the end of each foot, which greatly increased the robot's traction.

Another possible modification considered was the number of legs. If a robot has more legs, different gaits are possible, some of which might be more efficient and faster. Also, more legs gives more weight-bearing capability and less friction at each point where a leg touches the ground, which may result in faster speeds for higher loads. If a robot must carry a processor and a battery, eight feet might provide better traction and support the weight better.

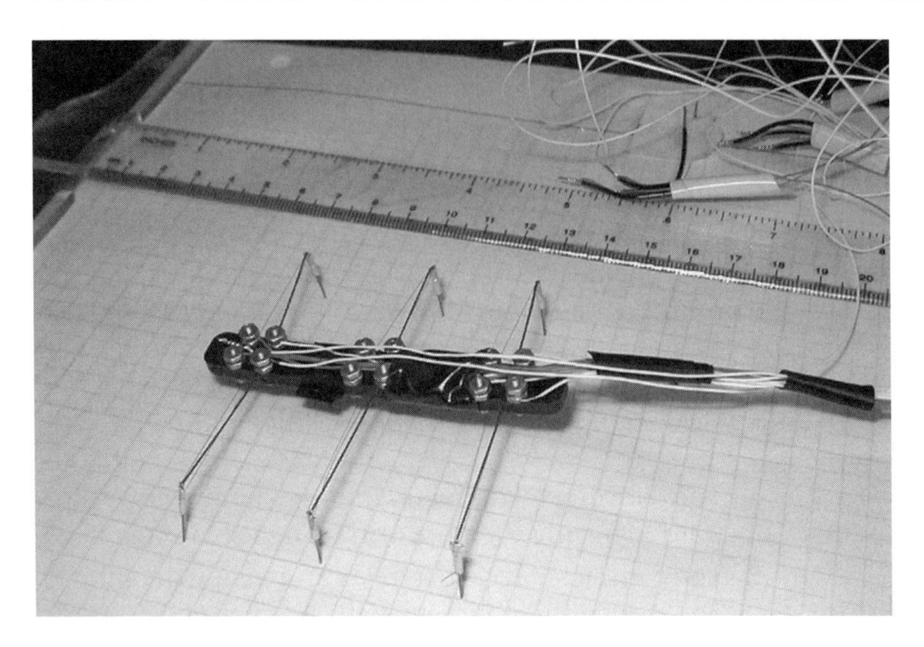

Figure 7.2 Robot B, a two-degrees-of-freedom Stiquito, built with screws.

Another possibility that can be considered is the length of the legs. With a longer leg length, a Stiquito can realize a greater arc of contraction, which pulls the robot farther forward with each step. The students were concerned about increasing the time to cool, since you would need more Flexinol®, but since the surface area also increases proportionally, they felt that the difference would be negligible.

A characteristic that the group found through experimentation was the way that the Flexinol® propelled the robot. There are at least two ways you can make the legs propel the rest of the robot: (a) the legs can push the robot forward as the Flexinol® contracts, or (b) the robot can be moved forward as the Flexinol® relaxes. This feature was tested with some prototype robots, as explained later in this chapter.

One factor that can only be controlled in software is the timing of the legs. Since the robot ran using the Parallax BASIC Stamp 2 processor to control the legs, students had to write software to control the robot. How efficiently the algorithms are designed and implemented in software can greatly influence variables such as how fast the leg contracts, how long it takes the leg to cool, or how long an overall stroke takes. All of these factors influence the speed of the robot.

Experimentation Details

In order to test all of the influencing factors, the group built a variety of robots, each exhibiting some modifications, and wrote software to control them. One robot built exactly according to the instructions in *Stiquito for Beginners* [1] was used as a control. This Stiquito had six legs, 30 mm long, each attached with crimps (Figure 7.1), and is called "Robot A."

The group then made several more test robots, as described below:

- Six-legged robot with two degrees of freedom (Robot B)
- Long-legged Stiquito (Robot C)
- Eight-legged robot (Robot D)

Stiquito With Two Degrees of Freedom (Robot B)

Materials for this robot (shown in Figure 7.2), beyond the standard Stiquito kit [1], include 12 screws, 24 nuts, 18 washers, and about 30 cm more Flexinol® than the basic Stiquito. This variant was built the same as the *Stiquito for Beginners* [1] kit instructions, with two exceptions: screws are used instead of crimps, and another piece of Flexinol® is used inside the crimps on the legs (the same length). The second piece of Flexinol® is the actuator for the height adjustment. For each leg, put another screw through the hole on the other side of the music wire, then put a washer over it, and then thread a nut on the screw. Wrap the Flexinol® around the screw between the nut and the washer. This will place the Flexinol® about 3⁄32″ higher than the Flexinol® already in place. The control wires attach in the same way for this set of actuators as the original leg screws.

Long-legged Stiquito (Robot C)

Materials for this robot, beyond the standard Stiquito kit [1], include 6 screws, 6 washers, 12 nuts, 60 cm of 0.032″ music wire (the kit comes with 0.020″), and about 30 cm more Flexinol® than the basic Stiquito. The main objective of this variation was to provide a larger step than the standard Stiquito. This robot uses three 200 mm lengths of 0.032″ music wire instead of 100 mm of 0.020″ music wire. It uses longer lengths of Flexinol® as well—approximately 60 mm for each leg. This robot was built with the basic instructions from *Stiquito for Beginners* [1], except it uses screws instead of crimps to attach the Flexinol® to the body [1, page 169].

There were two reasons for using thicker wire: the longer legs must be stiff enough to hold Stiquito off the ground at the body, and the larger music wire pulls the Flexinol® out with more force. This robot is shown in Figure 7.3.

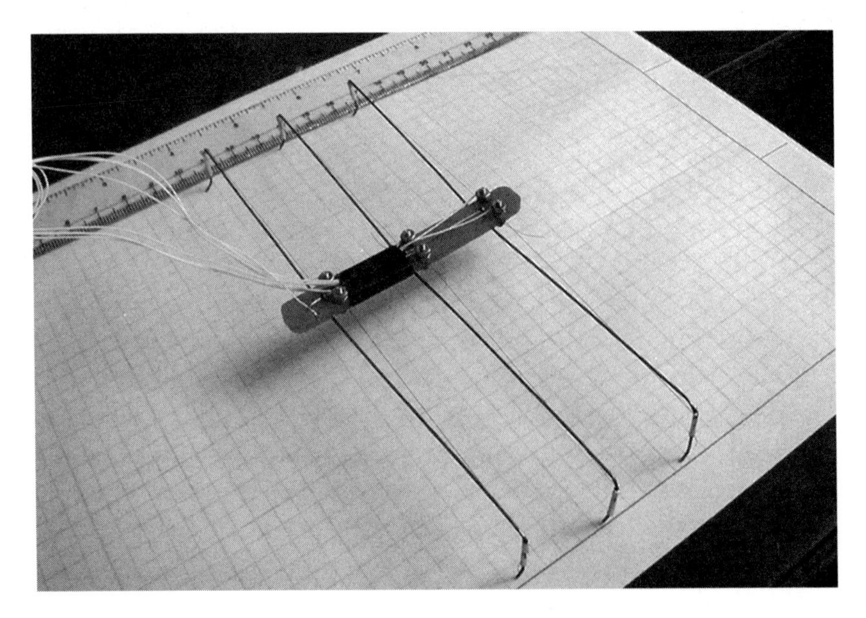

Figure 7.3 Robot C, long-legged Stiquito.

Eight-legged Stiquito (Robot D)

Materials for this robot, beyond the standard Stiquito kit [1], include 8 screws, 16 nuts, 8 washers, 10 cm more music wire, and 12 cm more Flexinol®. Since the standard Stiquito body has holes for only three pairs of legs, a new body was needed. The group used the plastic stick provided for the manual controller for the body, and drilled their own holes for the legs and screws. The leg holes were drilled as close to the ends of the body as possible, and the remaining two pairs of legs were placed so all legs were equidistant from each other. Other than having a fourth 100 mm music wire leg piece and more Flexinol®, the assembly of this one followed the book instructions [1]. This robot is shown in Figure 7.4.

Test Results

In order to test the robots, the students ran the four robots through twelve trials. Each trial measured the time it

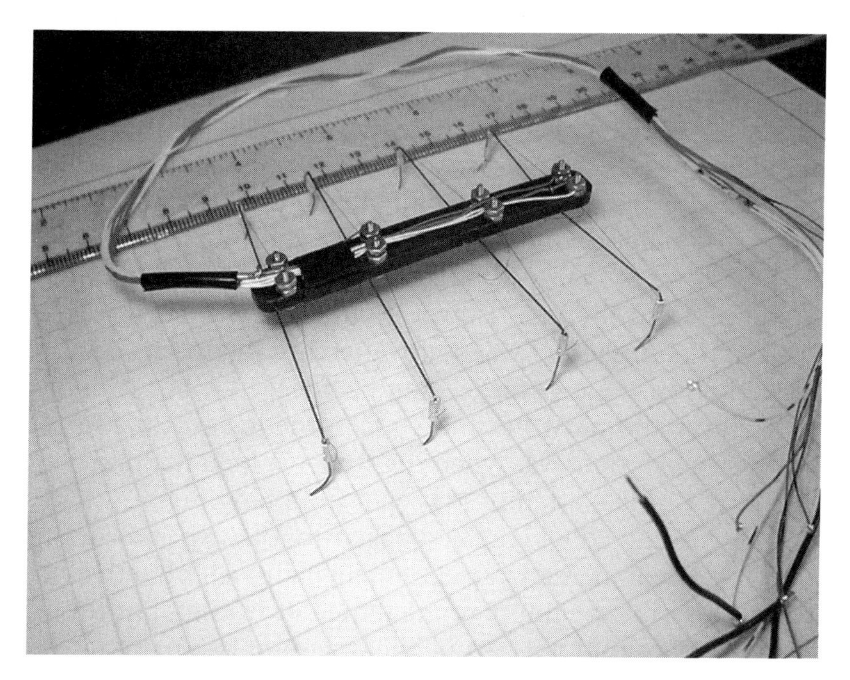

Figure 7.4 Robot D, eight-legged Stiquito.

took the robot to travel 5 cm. The robots were all carrying weight—an AAA battery wrapped in electrical tape. All robots were tested with the tripod gate, and all had ratchet feet except the two-degrees-of-freedom robot. The current for all of the robots was 600 mA at 4.2 V. The robot that won the race was carrying a board with a BASIC Stamp, a ULN chip, and a small 6 V (type 28a) battery to power the BASIC Stamp. The current for the legs came from an adjustable, off-board power supply, providing up to 2 A at 5 V.

The software for all of the robots was optimized for their particular hardware. Each version of code used pulse-width modulation, although the PWM timing for the long-legged robot and the two-degrees-of-freedom robot had to be modified. The tripod gate timing also differed depending on the robot, how many legs and wires there were, and how much power each wire used. The long-legged robot had to run using special code that provided more power to the wire, affecting the pulse-width modulation.

Analysis and Influencing Factors

The factors assessed included leg length, number of legs, degrees of freedom, gait motion, and the software used to animate the legs. The trials showed that the actual gait, the power provided to the Flexinol®, and the presence of a tether also made a difference in the speed of the robots.

The first set of trials suggested that the eight-legged robot would perform the best. However, after a few trials it was noticed that the tether was really affecting the robots, and the tether on the eight-legged and six-legged robots were the lightest and most flexible. In fact, the lowest time for the six-legged robot was when the students held the tether up in the air in an attempt to keep it from tangling.

The winning robot, the six-legged robot with two degrees of freedom, had an extremely lightweight tether. The group tried to eliminate the tether altogether, but the robot simply could not handle the weight of two batteries. Since the other robots did not have on-board processors, they had much heavier tethers, which slowed them down. The robot with the lightest tether, after the winning robot, was Robot A, the six-legged control robot. You will notice the moderately low time of 59 s for run two of this robot, which was after the tether was adjusted so the robot was not dragging any of the tether wire.

Final Implementation and Software Emphasis

Through this experimentation, the students determined that two degrees of freedom and longer legs would perform best if the walking algorithms were optimized. After the first set of trials, they assembled a six-legged robot with two degrees of freedom, and worked with the software. They finally determined that the amount of power supplied to the legs also made a noticeable difference. The legs need a strong power supply, and a bench power supply at 2 A and 5 V was used. This is also related to the source code—the PWM timings can affect how much power the Flexinol® gets. This robot ended up winning the race, eclipsing all others by walking 1 foot in 40 seconds.

Software was the second biggest factor influencing the speed of the robot. The standard code provided by the professor worked fairly well for the short-legged robots with one degree of freedom. For the long-legged robots, modifications to the algorithm and code were needed. The biggest change made was to provide more power for longer periods of time to the legs. The long-legged robot that did not have power for a sufficient amount of time had very poor trial times. The new algorithm also did not taper off the PWM, like the code provided did. Once again, this provided more power to the Flexinol®, reinforcing the idea that power to the Flexinol® is a major factor, whether it is controlled from the software or the hardware.

Other software-influenced factors are how fast the legs move, and (in the two-degrees-of-freedom robot) what the gait algorithm implemented. When the contraction and relaxation cycle of the robot's legs is shortened, the gait becomes much quicker. This is only limited by how long the Flexinol® takes to cool. For the two-degrees-of-freedom robot, the students spent a lot of time trying to figure out the best gait algorithm. They first tried up and forward simultaneously, then down, then relaxing back and pushing the robot forward. This did not work very well, as shown by the times in Table 7.1. Then they tried contraction (pushing the robot forward), up and relax, and then down. As the Flexinol® was cooling, it was also being pulled back into the long shape. This difference in the gait made a huge difference in speed.

The difference between the contract and the relax phases of the Flexinol® pushing the robot is surprising. Depending on how the feet are ratcheted, the Flexinol® will either push the robot forward when it contracts or when it is relaxed. Experimenting with the gait, the students noticed that it worked much better if the feet were ratcheted in such a way that they caught the ground and pushed the robot forward as the Flexinol® contracted. For two degrees of freedom, it helped even more that the up Flexinol® slightly opposed the forward Flexinol®. If the robot had up and horizontal firing at the same time, then the leg would not move at true 90° angles and it would not get the same

Table 7.1 Experimental data

Robot	Run 1	Run 2	Average
Robot A: Six-legged, one degree of freedom, weak power supply	90 s	59 s	74.5 s
Robot A: Six-legged, one degree of freedom, strong power supply	21 s	6 s	14 s
Robot B: Six-legged, two degrees of freedom, pushing forward on relax, weak power supply	79 s	77 s	78 s
Robot B: Six-legged, two degrees of freedom, pushing forward on contraction, strong power supply	6.7 s	10.8 s	8.7 s
Robot C: Six-legged, long-leg robot, weak power supply	79 s	89 s	84 s
Robot D: Eight-legged robot, weak power supply	72 s	63 s	67 s

strength of contraction as when the upward contraction is fired at a separate time from the forward contraction.

Experimentation on the final robot revealed that traction was a problem. Glue dots from a hot-glue gun were added to the feet to further enhance the robot's traction. One reason the eight-legged robot did so well was because it had extra traction from the extra feet. The two-degrees-of-freedom robot also had reduced friction since the other legs did not have to skid along forward as the Flexinol® pushed it forward. The winning robot is shown in Figure 7.5.

Conclusions and Future Work

Some factors to be assessed are the effects of using different gaits, weights, and types of construction. Further experimentation with the software is also needed. The screws did not hold the Flexinol® tight enough, and the amount of slack might have affected the results, but no concrete data was gathered. The tether had the biggest impact on the times for the robots without the on-board processor. Trials should be run with all the robots with a

Figure 7.5 The winning Stiquito robot.

processor on board, with a consistent power supply, and with a nonbinding tether to further prove the best design and implementation of a Stiquito variant.

References

[1] Conrad, J. M., and J. W. Mills, *Stiquito for Beginners: An Introduction to Robotics,* Computer Society Press: Los Alamitos, CA, 1999.

[2] Conrad, James M., and Jonathan W. Mills, *Stiquito: Advanced Experiments with a Simple and Inexpensive Robot,* IEEE Computer Society Press: Los Alamitos, CA, 1997.

Further Reading

Eskin, M. "Small Steps: An Experiment with a Small Robot," http://www.computer.org/books/stiquito/eskin.html, 1998.

Halliday, Resnik, and Walker. *Fundamentals of Physics,* 5th edition, Wiley, 1996.

Chapter 8

More Stiquitos Controlled

ISBN 0-471-48882-8 © 2005 IEEE Computer Society

Stiquitos have been built in all shapes, sizes, and variants. Many of our previous readers have written to us describing robots they have built. They have detailed their design concepts, hardware, software, and assembly processes. Many, like Luke Penrod of Texas A&M, have sent us pictures showing their new creations (Figure 8.1). This chapter shows some examples of Stiquito or Stiquito-like robots that have control circuitry.

Analog Controller

You can build a simple analog-based circuit board that mounts on top of an already built and tested Stiquito robot and directs the robot to walk with a tripod gait. This circuit feeds current to the Flexinol® on a periodic basis, which you can adjust. This setup allows Stiquito to be an autonomous robot. The book *Stiquito for Beginners: An Introduction to Robotics* [1] includes a schematic for such a circuit. The circuit generates enough current to contract the Flexinol® actuators, three wires at a time. It uses a pop-

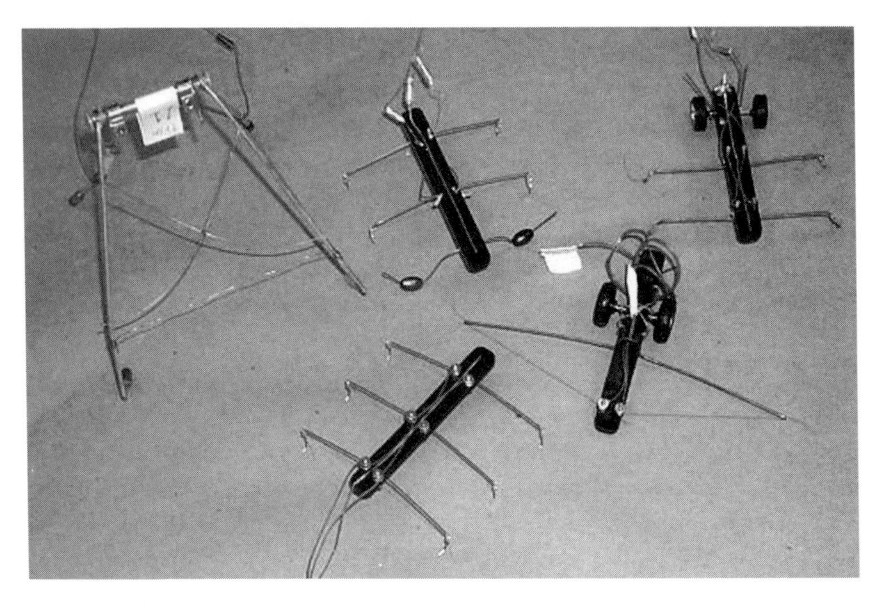

Figure 8.1 Some examples of Stiquito variants from Texas A&M.

ular timer circuit to generate a pulse to contract the legs. The two main parts of the circuit are:

1. The actuator components (LEDs, transistors, and Flexinol®) for user feedback and motion
2. The pulsing components (555 timer, capacitors, potentiometer, and resistors) for generating the pulses to the Flexinol® legs

Serge Caron modified the circuit to create a better-timed analog circuit [2]. The design provides a more current-efficient gait controller than the one described in the book *Stiquito for Beginners* [1]. The circuit uses a shift register to allow a tripod to relax before the other tripod starts contracting. He even experimented with using flip-flops instead of shift registers, since the CD4035 was difficult to obtain. A photograph of the completed robot is shown in Figure 8.2.

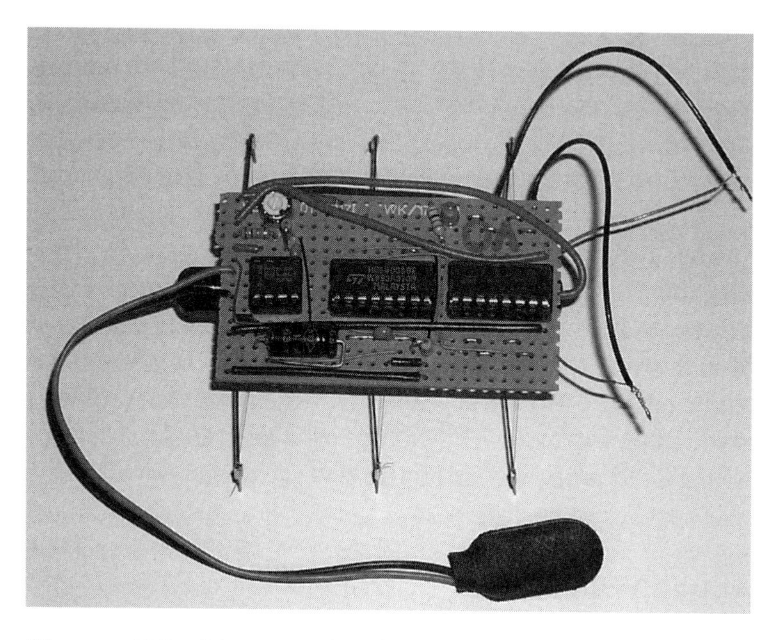

Figure 8.2 Photograph of the shift-register circuit attached to Stiquito.

BEAMStiquito

One example of a simple controller for Stiquito, named "BEAMStiquito," was built by Mark van Dijk of the Netherlands [3]. BEAM (biology, electronics, aesthetics, mechanics) is a concept of building robots that have a high survivability chance. These robots are often self-sustaining; they use solar power and try to minimize power consumption. The most common BEAM robots are Photovores, Solarollers, and Walkers [4]. Most BEAM robots resemble tiny insects. Characteristics of these designs include a low number of parts and the high flexibility of the circuit. These designs often use the concepts of neural networks, with which the system "learns" the optimal operation by training itself to generate specific outputs based on specific inputs.

Each leg uses a separate Nv neuron, so the six-legged Stiquito would need a nervous network of at least six neurons. This also means that it uses six inverters. Many BEAM robots employ circuits with HCT integrated circuits (ICs). The 74HCT14 was chosen because it employs a Schmitt trigger and runs at a low voltage. This IC also contains the six inverters needed for the Nv neurons that the BEAM pattern needs to drive current to the actuators. There is a resistor–capacitor network that separates each inverter. The delay of the RC network is one second for switching because the Flexinol® needs time to heat up and to cool down.

The dimension of the robot is 70 mm wide by 75 mm long by 50 mm high (see Figure 8.3). The weight of the robot is 50 grams, including batteries. Its speed is about 3–5 mm a step. It can step down from objects about 4 mm high. At the moment, the only direction the robot can walk in is forward. The layout of the hand-made circuit board used is shown in Figure 8.4.

Atmel AT90s8535 Controller

Cosmin Rotario built a Stiquito-like robot using Flexinol®, music wire, and perf board. He attached the robot to a

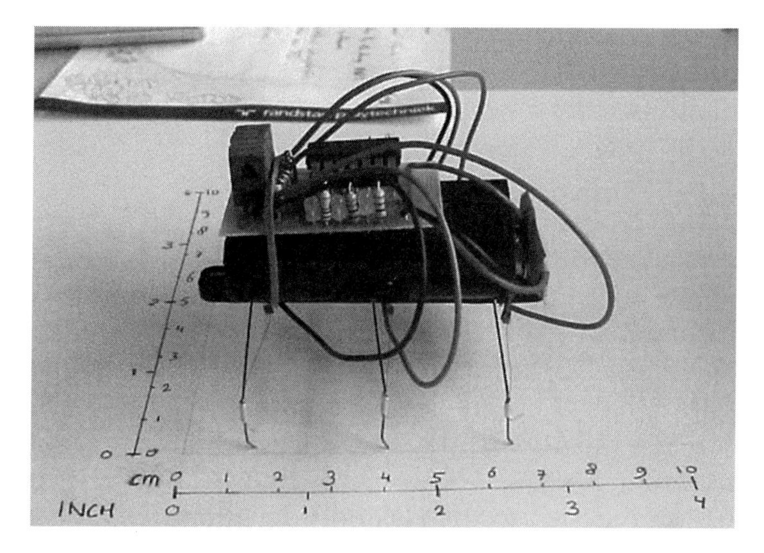

Figure 8.3 A fully functional BEAMStiquito. Notice that the battery is between the circuit board and the Stiquito robot.

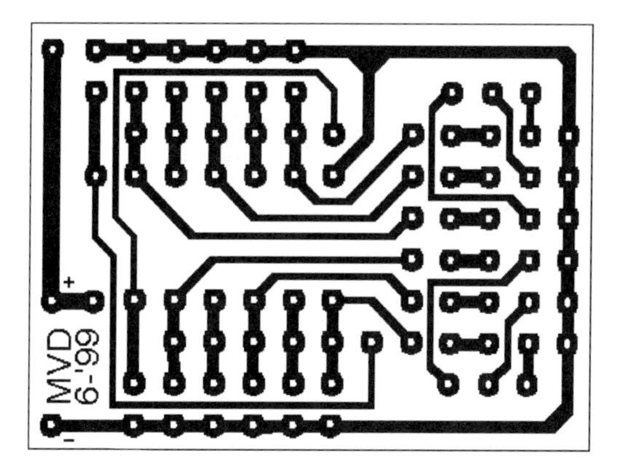

Figure 8.4 The printed circuit board (PCB) artwork for the BEAMStiquito. The solder side is shown.

full-functioned controller via a seven-wire tether. The controller is based on an Atmel AT90s8535 microprocessor and includes a keypad input and HD44780 LCD output. His system is shown in Figure 8.5.

Atmel AVR-Based Controller

Christian Lehner of Berlin, Germany, designed and built AVR Stiquito, a little hexapod robot based on Stiquito (Figure 8.6). The legs are controlled by a little 20 pin microcontroller, the Atmel 2313. The signals from the processor are amplified using a Darlington transistor array (ULN 2001) and are fed to the Flexinol® actuators, which contract when heated by the current flowing through them.

Mechanically, the PCB is similar to the original Stiquito's plastic body. There are three pairs of holes in the middle of the board, which hold the legs (Figure 8.7). In order for Stiquito to walk properly, it is important to mount the legs by inserting the piano wire from the bottom. That way, the legs stand above the PCB and the muscle wires pull them back and down to lift and move the tiny crea-

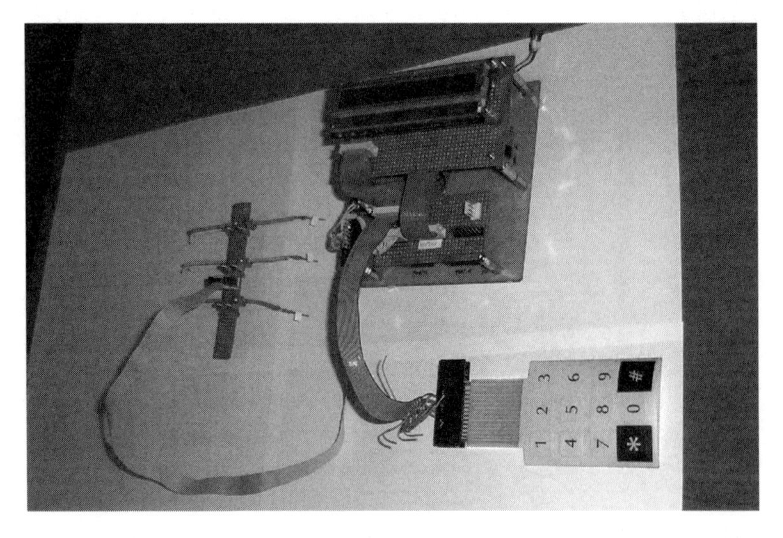

Figure 8.5 An Atmel AT90s8538-based Stiquito controller.

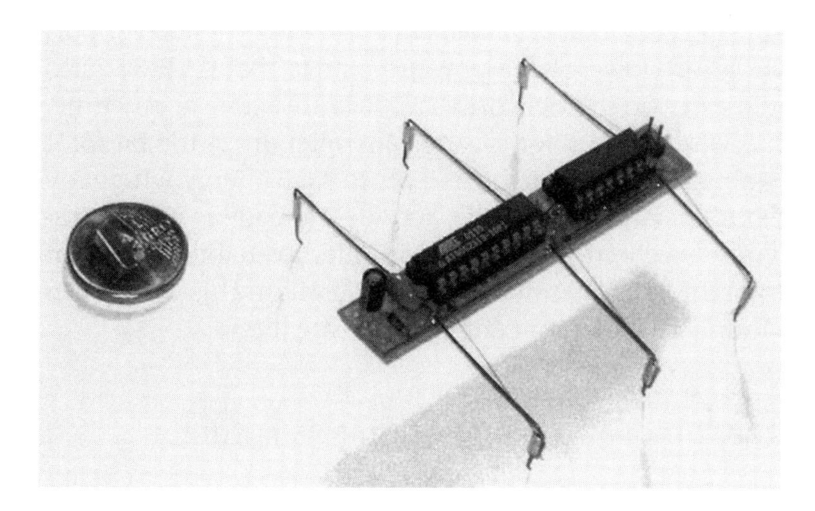

Figure 8.6 AVR Stiquito, without the power tether.

ture. One important step in building this robot is the attachment of the Flexinol® wires at the PCB body side. The wires should be inserted from the top. Electrically, the pull from the leg is sufficient to guarantee contact. A knot at the other end of the wire is used to hold the wire in the aluminum tube attached to the knee of each leg. In order to apply force to the wire to return it to its relaxed state, the piano wires should be bent forward at a 45° angle and be pulled back to 0° (or 90° relative to the body) for attachment of the actuators at the leg.

The software generates pulse-width-modulated signals for the Flexinol® wires. The software also toggles the state

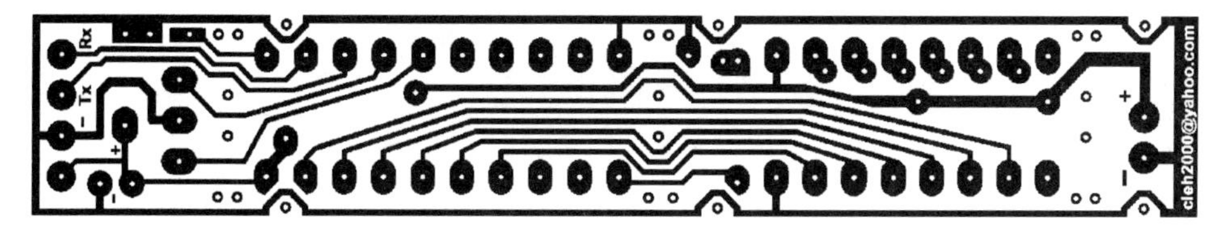

Figure 8.7 Circuit board layout of the AVR Stiquito, which uses the Atmel AT90S2313 processor (bottom layer).

of the LED, simulating a heartbeat. In order to make it possible to use low-voltage power supplies that AVR Stiquito could carry on its back, only two legs are active at any time. Otherwise, the voltage for the controller drops too far for it to operate. The LED is turned off to save energy when any Flexinol® actuator becomes active. The code is written using the excellent AVR-GNU compiler available for free at the AVR Freaks Website [5]. More detail on this design can be found at Christian Lehner's Website [6].

Other Innovations and Stiquito's Future

Students have created many non-Stiquito robots using Flexinol®. Many designs were innovative, and some did not work. What better way to test your design than to build an inexpensive Flexinol®-propelled robot? Three designs are shown in Figures 8.8 and 8.9. Figure 8.8 shows a caterpillar-like design, which did not quite work as hoped. Figure 8.9 shows two successful walkers. The robot on the left uses sticks and an old CD (cut in half) to reach forward

Figure 8.8 A "caterpillar" Stiquito design.

Figure 8.9 Two walking Flexinol® robots racing each other.

and pull itself forward, similar to how one would walk with crutches. The one on the right used the Flexinol® to pull the weight of the robot along rollers, and "scooted" along the table surface when the Flexinol® was relaxed.

Over the years, high school, community college, and university faculty have used Stiquito to educate tomorrow's engineers. Texas A&M, New Jersey Institute of Technology, Penn State, and NC State University are just a few of the larger programs to use Stiquito. Scores of junior high and high school students choose Stiquito for their science fair projects. Thousands of hobbyists choose Stiquito to dabble in robotics. Why Stiquito?

Some of the last words in *Stiquito for Beginners* [1] still ring true today:

> A colleague of Jonathan Mills observed that the longevity of Stiquito is probably due to two things: it is simple, and everybody believes s/he can build a better one!
>
> We agree. As an inexpensive introduction to robotics, this little 'bot can get you started. But you do not have to stop there. So why build Stiquitos? We build them because they are simple, they are fun, and they let us investigate

robotics without getting a million-dollar grant. We encourage all readers of this book to improve on Stiquito. If you build a better yet inexpensive ant-like robot, we will stand in line to get one!

The excitement generated by Stiquito still continues more than 13 years after its introduction. We hope that we have continued to "kindle the robotics spirit" in many of you. Should you build a better Stiquito, don't forget to let us know. We plan to publish a fourth Stiquito book showcasing your innovative designs.

References

[1] Conrad, J. M., and J. W. Mills, *Stiquito for Beginners: An Introduction to Robotics,* IEEE Computer Society Press: Los Alamitos, CA, 1999.

[2] Conrad, J. M., and S. Caron, "A Simple Circuit to Make Stiquito Walk on Its Own Effectively," *Robot Science and Technology Magazine,* pp. 14–19, Issue 8, Feb./Mar. 2001.

[3] Conrad, J. M., and M. van Dijk, "BeamStiquito," *Circuit Cellar Ink,* pp. 10–12, Issue 120, July 2000.

[4] Trachtman, P., "Redefining Robots," *Smithsonian,* February 2000, pp. 96–112.

[5] AVR Freaks website: http://www.avrfreaks.net/.

[6] AVR Stiquito website: http://www.geocities.com/cleh2000/robots/stiquito/.

Appendix

Sources of Materials for Stiquito

This appendix lists some suppliers of Stiquito parts, robotics kits, and electronics. You may find many Stiquito supplies and tools at your local hobby supply store. Check local electronics supply stores for other supplies. This section is not an endorsement of these companies, but is provided to make your supplier search easier.

All Electronics Corporation
P.O. Box 567
Van Nuys, CA 91408
Phone 800-826-5432, 818-904-0524; Fax 818-781-2653
E-mail allcorp@allcorp.com
Web address http://www.allelectronics.com
Surplus dealer of boards, components, and assemblies.

Artificial Creatures
22 McGrath Hwy., Ste. 6
Somerville, MA 02143
Phone 617-629-0055; Fax 617-629-0126
E-mail art@isr.com
Web address http://www.isr.com
A subsidiary of IS Robotics. Supplies small mobile robots for research and education.

ISBN 0-471-48882-8 © 2005 IEEE Computer Society

Digi-Key
701 Brooks Ave. South
P.O. Box 677
Thief River Falls, MN 56701-0677
Phone 800-344-4539 or 218-681-6674; Fax 218-681-3380
Web address http://www.digikey.com
Supplier of electronic components and other supplies.

Dynalloy, Inc.
3194-A Airport Loop Drive
Costa Mesa, CA 92626-3405
Phone: 714 436-1206, Fax: 714 436-0511
E-mail sales@dynalloy.com
Web address http://www.dynalloy.com
Supplier of nitinol wire, trade named Flexinol®. Flexinol® can be ordered in lengths of
1 meter or more.

Edmund Scientific
60 Pearce Ave
Tonawanda, NY, 14150
Phone 800-728-6999; Fax 800-828-3299
Web address http://www.scientificsonline.com
Sells optical components, science kits, surplus motors, and robot kits.

Express PCB
E-mail support@expresspcb.com
Web address http:// www.expresspcb.com
They provide a free circuit design and layout software package. The output from the package can be used by them to make a few, very affordable printed circuit boards.

iRobot
63 South Aveue
Burlington, MA 01803
Phone 781-345-0200; Fax 781-345-0201
E-mail info@irobot.com
Web address http://www.isr.com
Source for research robots and sensor systems.

Jameco Electronics
1355 Shoreway Rd.
Belmont, CA 94002-4100
Phone 800-831-4242 or 415-592-8097; Fax 800-237-6948
E-mail purchasing@james-electronics.com
or international@jameco.com
Web address http://www.jameco.com
Supplier of electronic components.

K&S Engineering
6917 W. 59th St.
Chicago, IL 60638
Phone 773-586-8503; Fax 773-586-8556
E-mail ksmetals2@aol.com
Web address http://www.ksmetals.com
Supplier of music wire and aluminum, copper, and brass tubing. Minimum order of $80.00. (Hobby shops also carry these items.)

LEGO DACTA
Lego Educational Division
113 North Maple Street
P.O. Box 1600
Enfield, CT 06083
Phone for education purchases: 860-763-3211
Phone for retail purchases:800-673-0360
Web address http://www.lego.com
Sells components needed for quickly building robot prototypes; educational department sells primarily to schools.

Micro Fasteners
24 Cokesbury Rd, Suite 2
Lebanon, NJ 08833
Phone 800-892-6917 or 908-236-8120; Fax 908-236-8721
E-mail info@microfasteners.com
Web address http://www. microfasteners.com
Supplier of brass #0 screws, nuts, and washers used on the Stiquito body.

Micro Robotics Supply, Inc.
14922 Northgreen Drive
Huntersville, NC 28078
E-mail orders@stiquito.com
Web address http://www. stiquito.com
Supplier of the Stiquito Repair Kit, Screws Kit, analog controller, parallel controller, and other Stiquito accessories.

Mondo-tronics, Inc.
124 Paul Drive, Suite 12
San Rafael, CA 94903
Phone 800-374-5764 or 415-491-4600; Fax 415-491-4696
E-mail info@RobotStore.com
Web address http://www. RobotStore.com
Supplier of nitinol and Flexinol® wire, robots, robotic books, videotapes, even robotic artwork.

Mouser Electronics, Inc.
1000 North Main Street
Mansfield, TX 76063-4827
Phone 800-346-6873 or 817-804-3888; Fax 817-804-3899
E-mail sales@mouser.com
Web address http://www. mouser.com
Wide selection of electronic components. Regional distribution centers; will fax detailed specifications. Accepts small orders.

Newark In One
4801 N. Ravenswood
Chicago, IL 60640-4496
Phone 800-463-9275, 773-784-5100; Fax: 888-551-4801
Web address http://www.newark.com
Distributor of electronic components.

Parallax, Inc.
599 Menlo Drive, Suite 100
Rocklin, CA 95765
Phone 888-512-1024; Fax 916-624-8003
E-mail: sales@parallax.com
http://ww.parallax.com, http://www.stampsinclass.com
Supplier of the BASIC Stamp microcontroller, educational curriculum, and robotics kits.

Parts Express
725 Pleasant Valley Drive
Springboro, OH 45066-1158
Phone 800-338-0531 or 937-743-3000; Fax 937-743-1677
E-mail xpress@parts-express.com
Web address http:// www.parts-express.com
Supplier of electronics, tools, hardware, and supplies.

Plastruct, Inc.
1020 S. Wallace Pl.
City of Industry, CA 91748
Phone 626-912-7016 or 800-666-7015; Fax 626-965-2036
E-mail plastruct@plastruct.com
Web address http://www.plastruct.com
Supplier of plastic stock used in Stiquito II, Boris, and SCORPIO. Plastruct offers a 30% educational discount to instructors and schools using a purchase order.

Radio Shack
Phone 800-843-7422
Web address http://www.radioshack.com
National chain; consult telephone directory for nearest dealer. Offers a variety of electronic components from local distributors.

Small Parts, Inc.
13980 NW 58th Court
P.O. Box 4650
Miami Lakes, FL 33014-0650
Phone 800-220-4242 or 305-558-1255; Fax 800-423-9009 or 305-821-8930
E-mail parts@smallparts.com
Web address http://www.smallparts.com
Supplier of metal, plastics, tools, and hardware.

Solarbotics Ltd.
179 Harvest Glen Way NE
Calgary, Alberta
Canada T3K 4J4
Phone 866-276-2687; Fax/alternate 403-226-3741
Web address http://www.solarbotics.com
A source for solar-powered robot kits.

Texas Instruments
Web addresses http://www.ti.com and
http://www.ti.com/home_p_micro430
The TI MSP430 microcontroller is the brains of Stiquito. Their site contains the free development tools described in this book. Look on the Web pages for Design Resources, then Development Tools. Download the IAR Embedded Workbench Tools.

Glossary

Actuator: A device responsible for moving a mechanical device, such as one connected to a computer by a sensor link.

A/D conversion (ADC): Analog-to-digital conversion is an electronic process in which a continuously variable (analog) signal is changed, without altering its essential content, into a multilevel (digital) signal.

Algorithm: A finite set of step-by-step instructions that can be followed to perform a specific task, such as a mathematical formula or a set of instructions in a computer program.

Ambient temperature: The temperature of the air that surrounds operating equipment.

Analog: Circuitry or components exhibiting continuously variable (nondiscrete) properties.

Autoroute: A computer-aided process of routing the PCB traces. The autoroute algorithm requires the schematic netlist to have the information about the component connections.

Autonomous: Not controlled by others or by outside forces; independent.

AWG: American Wire Gauge. A number in front of

ISBN 0-471-48882-8

"AWG" specifies a thickness of the wire; the smaller the number, the thicker the wire.

BASIC: Beginners All-purpose Symbolic Instruction Code. BASIC is a widely used, high-level programming language.

CAD: Computer-aided design. CAD is a general term that includes any software that aids in the design of products.

Capacitor: An electric circuit element used to store charge temporarily. It consists, in general, of two metallic plates separated and insulated from each other by a dielectric.

CD (compact disc): A small optical disk on which data such as music, text, or graphic images is digitally encoded.

Circuit: A configuration of electrically or electromagnetically connected components or devices.

Compiler: A program that translates another program written in a high-level language into machine language so that it can be executed on a computer.

Component: An integrated circuit.

CPU (central processing unit): An older term for processor and microprocessor, the central unit in a computer containing the logic circuitry that performs the instructions of a computer's programs.

Crimp: To pinch or press together in order to seal or fasten; also, the mechanical attachment of Flexinol® to the Stiquito robot leg using compressed aluminum tubing.

Current: The amount of electric charge flowing past a specified circuit point per unit time.

Darlington sink driver: see ULN2803A.

Debug: The process of detecting, locating, and correcting a problem in a software program or hardware.

Degrees of freedom: Specifies the number of independent movements that an appendage can make (example: legs that can move backward and can be raised). Modern manufacturing robots have six degrees of freedom.

Digital: Digital describes electronic technology that generates, stores, and processes data in terms of two states: positive and nonpositive. Positive is expressed or represented by the number 1 and nonpositive by the number 0.

DIP: Dual in-line package. An electronic chip or other component that has two rows of "legs" that are placed in a printed circuit board through holes in the board.

DIP switch: Dual in-line package switch. Printed circuit boards and peripheral devices are often equipped with a bank of DIP switches to control various aspects of the board's operation.

Double-sided board: A printed circuit board that has traces on both sides.

DRC (design rule check): The DRC is a computer routine that checks against a design error list compiled by the user. DRCs mostly check for spacing and sizing violations and check for unfinished connections.

EEPROM: Electrically erasable programmable read-only memory. A memory chip that maintains data content after power has been removed. EEPROM is programmed and erased electrically.

EPROM: Erasable programmable read-only memory. A memory chip that maintains data content after power has been removed. EPROM is programmed electrically. It can be erased and reprogrammed, but can only be erased by shining ultraviolet light into the "window" on top of the device.

Embedded system: A commercial or other device that contains a microprocessor or microcontroller purchased as a part of the equipment. Examples include mobile phones, microwave ovens, and automobiles.

Fan-out: Routing small traces from the surface-mount component pads to a space where a via can be inserted to connect to the plane layers. This term is not to be confused with the digital device's definition of fan-out.

Flash memory: A type of constantly powered nonvolatile memory that can be erased and reprogrammed in units of memory called blocks. Sometimes called "flash read-only memory."

Flexinol®: A form of nitinol that has been processed to contract when heated and return to its original shape when cooled and stretched by a recovery force (i.e., spring wire). Flexinol® is a registered trademark and is produced by Dynalloy, Inc.

Flip-flop: An electronic circuit or mechanical device capable of assuming either of two stable states; especially a computer circuit used to store a single bit of information.

Flux: A chemical that aids in the soldering process. It cleans oxides from the pads and breaks the surface tension of solder.

Footprint: The footprint is the outline and pad location required by a component.

Gait: A particular way or manner of moving legs.

Hexapod: Having six feet.

Humanoid: Having a human form or characteristics.

IC: Integrated circuit. A complex set of electronic components and their interconnections that are etched or imprinted on a chip.

IDE (integrated development environment): A system for supporting the process of writing and testing software. Such a system may include a syntax-directed editor, graphical tools for program entry, and integrated support for compiling and running the program and relating compilation errors back to the source.

Interrupt: An asynchronous event that suspends normal processing and temporarily diverts the flow of control through an "interrupt handler" routine (or ISR). Interrupts may be caused by both hardware (I/O, timer, machine check) and software (supervisor, system call, or trap instruction).

Interrupt service routine (ISR): A subroutine that is executed when an interrupt occurs. Interrupt handlers typically deal with low-level events in the hardware of a computer system, such as a character arriving at a serial port or a tick of a real-time clock.

JTAG (Joint Test Action Group): 1. Committee that established the test access port (TAP) and boundary-scan architecture defined in IEEE Standard 1149.1-1990. 2. The common name for IEEE Standard 1149.1-1990 hardware and interfacing.

LCD (liquid crystal display): A digital display that uses liquid crystal cells that change reflectivity in an applied electric field; used for portable computer displays and watches.

LED (light-emitting diode): A semiconductor that emits light when electric current runs through it in one direction.

Locomotion: The ability to move from one place to another.

Microcontroller: A single-chip microcomputer with onboard program memory and input/output hardware that can be programmed for various control functions.

Microprocessor: A silicon chip containing a dedicated central processing unit (CPU). Usually used as the foundation for higher-order systems and circuits.

Multilayer board: A printed circuit board that has more that two layers.

Nitinol: An alloy of nickel and titanium that can have a number of properties including the shape memory effect, super elasticity, corrosion resistance, and actuation. Stiquito uses a specific form of nitinol called Flexinol®. See "Flexinol" for more information.

Pad: The copper connection (usually tinned) of a trace to which a component's pins are soldered.

Padstack: The detailed information about how the pad is made on different layers of a printed circuit board.

Parallax BASIC Stamp 2 processor: A single-board computer that runs the Parallax PBASIC language interpreter in its microcontroller. The developer's code is stored in an EEPROM, which can also be used for data storage. The PBASIC language has easy-to-use commands for basic I/O, like turning devices on or off, interfacing with sensors, and so on. More advanced commands let the BASIC Stamp interface with other integrated circuits, communicate with each other, and operate in networks. The BASIC Stamp has been widely used by hobbyists, in lower-volume engineering projects, and in education due to its ease of use and wide support base of free application resources.

Parallel port: An interface on a computer that supports transmission of multiple bits at the same time; used almost exclusively for connecting a printer. On IBM or compatible computers, the parallel port uses a 25-pin

D-shell connector. Macintoshes have an SCSI port that is parallel, but more flexible in the type of devices it can support.

PCB (printed circuit board): A flat piece of material, often fiberglass, covered with small, flat wires. PCBs are the means by which integrated circuit components are connected.

Perf board, perforated board: A type of circuit board that has only holes. This board is used for prototyping circuits.

Plane layer: A general term for a power or ground layer.

Platform: A hardware and/or software framework for experimentation.

Pneumatic: Using air pressure to move or activate a device.

Potentiometer (POT): A variable-resistance device, with the resistance value determined by a mechanical screw or wiper that is moved with a screwdriver or similar device.

Power bus: The 20-AWG copper wire that runs the length of the Stiquito robot body and provides voltage to the legs. (Also called spine.)

Propulsion: The action or process of moving something forward or backward

Prototype: One of the first units manufactured of a product. It is tested so that the design can be changed if necessary before the product is manufactured commercially.

Pulse-width modulation (PWM): The signal produced by PWM consists of a continuous pulse train of constant frequency and duration, based on the duty cycle. The amplitude of the pulses is constant.

RAM (random access memory): A temporary storage space in a computer where the operating system, application programs, and data in current use are kept so that they can be quickly accessed by the computer's processor. This storage is lost when the computer is powered off.

Reference designator: A label used to identify individual parts. The reference designator usually has a letter identifying the type of part and a number for the instance of the part.

Resistor: A device used to control current in an electric circuit by providing resistance.

Robotics: Study of robots.

ROM (read-only memory): Permanent memory storage for a computer; usually holds the software program by which the computer executes. ROM cannot be changed once it is manufactured.

Schematic: A structural or procedural diagram, especially of an electrical or mechanical system.

Sensor: A device that responds to a physical stimulus (such as heat, light, sound, pressure, magnetism, or a particular motion) and transmits a resulting impulse (for measurement or operating a control).

Silkscreen: The ink printed on a printed circuit board that gives the reference designators and other information about the parts or circuit board.

Single-sided board: A printed circuit board that has traces only on one side of the board.

Socket: A receptacle (holder) into which an electric device is inserted.

Software: The programs, routines, and symbolic languages that control the functioning of the hardware and direct its operation.

Solder mask: The solder-resistant lacquer that is placed on the board to prevent solder flowing where it is unwanted.

Solder side: The solder side of the board refers to the bottom layer. This is the side that the solder wave touches.

Soldering: A method of joining metals by using any of various fusible alloys and applying heat.

Spine: See Power bus.

State diagram: A diagram that shows the states of an object and the events that cause the object to change from one state to another.

Stiquito: A self-sufficient small, six-legged walking robot.

Thermal relief: A contact that connects a trace to a plane.

Timer: This module keeps track of time. If the selected

time interval expires, it generates an interrupt and executes any ISR associated with that timer.

Trace: A thin copper wire that is laminated onto the printed circuit board material.

Transistor: A small electronic device containing a semiconductor and having at least three electrical contacts; used in a circuit as an amplifier, detector, or switch.

Tripod: Device with three legs.

ULN2803A: A monolithic high-voltage, high-current Darlington transistor array. The device consists of eight NPN Darlington pairs that feature high-voltage outputs.

Via: A contact connecting a trace on one side of the printed circuit board to a trace on the other side.

Volatile: In the electronics realm, describes the ability of a memory device to retain its contents once the device is powered off. RAM is a volatile device; ROM, EPROM, EEPROM, and flash are nonvolatile devices.

Voltmeter: A device that measures voltage across a circuit.

Watchdog timer (WDT): The primary function of the WDT module is to perform a controlled system restart after a software problem occurs. If the selected time interval expires, a system reset is generated. For some devices, The WDT can also be configured as a timer to periodically trigger interrupts.

Index

555 timer, 161

A/D conversion, 25, 123, 127, 175
Actuator, 83, 92, 93, 161, 162, 164, 165, 166, 175
Advanced experiments, 23, 66
Aibo, 5
Algorithm, 19, 23, 39, 45, 47, 123, 139, 151, 155, 156, 175
Ambient temperature, 142, 175
Analog controller, 160
Analog-to-digital conversion, 17, 175
Asimov, Isaac, 3
Assembly skills, 70
Atmel 2313, 164
Atmel AT90s8535, 162, 164
Austenite, 65
Autonomous, 4, 5, 66, 160, 175
Autoroute, 45, 46, 47, 175
AWG, 175

BASIC, 176
BASIC Stamp, 130, 133, 134, 135, 136, 143
BEAMStiquito, 162, 163
Bill of materials, 39
Binary digit or bit, 20
Build Stiquito, 66
Byte, 20

C language, 116, 133
CAD, 37, 39, 40, 41, 44, 45, 47, 60, 176
CAD schematic capture, 37, 38, 40, 41, 44
Capacitor, 27, 41, 43, 176
Caterpillar, 64, 166
CD, 166, 176
Circuit, 176
Compiler, 176
Component, 176
Computer engineer, 17
Constructing Stiquito, 74–109
  assembling the legs, 78
  attaching leg crimps, 87
  checking alignment, 85
  cutting Flexinol®, 86
  leg crimps, 72, 86
  molded body, 75
  operating the Stiquito robot, 97
  parts list, 67, 68
  power bus, 66, 67, 74, 76, 79, 80, 82, 86, 95, 99, 180

ISBN 0-471-48882-8 © 2005 IEEE Computer Society

Constructing Stiquito *(continued)*
power bus supply, 95
preparing the screws on the body, 84, 85
ratchet feet, 82, 83, 94, 99, 154
securing the controller board to the robot, 96
stack-up of screws, 84
Stiquito assemblies, 74
tools needed, 67, 69
troubleshooting, 99
tying and sanding retaining knot, 87
CPU (central processing unit), 176
Crimp, 2, 9, 67, 70, 71, 72, 73
Current, 23, 26, 27, 43, 45, 50, 64, 66, 76, 92, 118, 131, 132, 134, 135, 136, 140, 142, 154, 160, 162, 164, 176
Cutting, 2, 9, 53, 67, 70, 86

Debug, 36, 116, 135, 176
Deburring, 70
Definition of a robotic device, 3
Degrees of freedom, 7, 176
Design rule check, 46, 47, 177
Design rule check (DRC), 39
Digital, 176
DIP switch, 135, 177
Double-sided board, 34, 48, 177
Dual in-line package (DIP), 34, 177
Dynalloy, Inc., 7, 170, 178

EEPROM, 17, 22, 133, 135, 177
Examples of uses of Stiquito, 2

Fan-out, 177
Flash memory, 25, 26, 113, 127, 177
Flexinol®, 6, 7, 9, 10, 21, 22, 23, 27, 64, 65, 66, 70, 71, 72, 73, 74, 75, 76, 83, 86, 87, 88, 89, 91, 92, 96, 98, 99, 100, 112, 118, 130, 131, 132, 134, 135, 136, 137, 138, 140, 141, 142, 147, 148, 149, 150, 151, 152, 153, 154, 155, 156, 157, 160, 161, 162, 164, 165, 166, 167, 177
crimping, 71
knotting, 71
relaxation time, 142
Flip-flop, 161, 178
Flux, 52, 56, 178
Footprint, 35, 40, 41, 42, 43, 44, 45, 47, 60, 178

Gait, 10, 11, 21, 22, 23, 25, 37, 64, 127, 132, 142, 147, 148, 149, 154, 155, 156, 157, 160, 161, 178
Ground plane, 34, 35, 45, 46

Hexadecimal, 20
Hexapod, 2, 7, 164, 178
Hobby-building skills, 2, 9, 70
Humanoid, 4, 178

IAR Embedded Workbench IDE, 116
i-Cybie, 5, 6
IDE (integrated development environment), 116, 178
Industrial arm robot, 4
Integrated circuits (ICs), 162, 178
Interrupt, 21, 119, 178
Interrupt service routine (ISR), 122, 178
Introduction to bioengineering, 10
Introduction to robotics, 10, 147, 167

JTAG, 23
JTAG cable, 103, 116, 117, 118
JTAG connection, 27, 103
JTAG connector, 27, 116, 118
JTAG port, 26, 113

K&S Engineering, 171

LCD (liquid crystal display), 164, 178
LED (light-emitting diode), 21, 22,

27, 57, 97, 99, 112, 141, 161, 166, 179
Locomotion, 23, 130, 179

Manufacturing and industrial engineers, 18
Martensite, 64, 65
Micro Fasteners, 171
Micro Robotics Supply, 139, 171
Microcontroller, 2, 11, 16, 17, 19, 20, 21, 22, 23, 25, 26, 27, 29, 30, 37, 38, 67, 70, 99, 103, 111, 112, 113, 116, 118, 127, 130, 133, 135, 141, 164, 179
Microprocessor, 11, 16, 17, 19, 130, 131, 132, 133, 179
Mobile phone, 16, 17, 49, 50
Multilayer board, 34, 43, 45, 53, 179

Netlist, 39, 40, 41, 42, 44
Nitinol, 7, 64, 177, 179

One degree of freedom, 7, 70, 103, 123, 126, 155
OrCAD, 41

Pads, 35, 40, 41, 42, 44, 45, 52, 54, 56, 105, 106, 108
Padstack, 40, 41, 51, 179
Parallax Basic Stamp, 23, 148, 150, 179
Parallax PBASIC language, 135, 179
Parallax, Inc., 143, 172
Parallel port, 27, 29, 103, 116, 148, 179
Photoresist, 50, 51
Plane layer, 34, 46, 180
Platform, 6, 10, 23, 64, 180
Pneumatic, 5, 180
Potentiometer, 21, 22, 27
Power, 21, 27, 29, 34, 36, 38, 42
Powerbus (*see* Constructing Stiquito)
Printed circuit board (PCB), 10, 11, 22, 23, 36, 74, 95, 103, 113, 133, 135, 163, 181
- artwork, 39, 48, 49, 50, 51, 54, 163
- drilling, 49, 53
- Gerber, 47, 48, 51
- laminate, 49, 50
- layout, 27, 29, 33, 36, 38, 39, 40, 41, 42
- manufacturing, 25, 33, 35, 41, 47, 48, 49, 58, 59, 60
- placement, 29, 42, 47
- populating, 54
- routing, 30, 34, 40, 43, 45, 46, 47
- silk screen layer, 47
- solder mask, 22, 47, 51, 52, 181
- stencil, 48, 49, 51, 54, 60
- testing, 2, 36, 56

Problem solving, 18
Product development, 18
Programming the MSP430, 116, 119
Propulsion, 5, 6, 7, 130, 180
Prototype, 23, 26
Pulse-width modulation, 112, 136, 154, 180
Pulse-width modulation (PWM), 115

Random access memory (RAM), 17, 22, 26, 133, 180
Read-only memory (ROM), 17, 177, 181
Reference designator, 39, 40, 47, 54, 180
Reset, 29
Reset switch, 29, 103
Resistor, 27, 40, 41, 54, 57, 162, 181
Robot, 1, 2, 3
Robotic arm, 4
Robotics, 1, 2, 3, 10, 11, 63, 127, 167, 181

Safety practices, 12
Sanding, 2, 9, 70, 71, 87, 140
Schematic, 27, 28, 36, 37, 38, 39, 40, 41, 44, 60, 133, 134, 135, 160, 181
Schematic capture, 19, 60

Science fair project, 167
Scientists, 17
Senior Capstone projects, 10
Sensor, 10, 23, 25, 64, 101, 127, 135, 139, 142, 181
Shift register, 161
Silkscreen, 181
Single-sided board, 34, 181
Socket, 46, 133, 135, 139, 181
Software, 5, 16, 17, 19, 21, 23, 25, 29, 39, 61, 66, 99, 103, 115, 116, 118, 119, 123, 127, 131, 132, 135, 136, 137, 139, 140, 142, 143, 147, 150, 151, 154, 155, 157, 160, 165, 181
Software architecture, 119, 120
Software engineers, 17
Solder mask, 51, 56, 181
Solder side, 52, 163, 181
Soldering, 9, 34, 56, 70, 71, 103, 104, 105, 106, 107, 108, 109, 181
Spine, 38, 41, 180, 181
State diagram, 115, 124, 126, 137, 138, 142, 181
Sticky robot, 7, 11, 64
Stiquito controller board, 11, 15, 19, 21, 22, 23, 24, 25, 26, 27, 28, 29, 30, 55, 56, 57, 58, 59, 61, 68, 70, 85, 96, 103, 107, 108, 109, 116, 117, 118, 119, 127
  prototype, 24, 25
  requirements, 21
  schematic, 36
Stiquito for Beginners, 9, 41, 66, 151, 152, 160, 161, 167
Stiquito II, 8, 11
Stiquito: Advanced Experiments, 9
Systems engineers, 18

Test fixture, 56, 57, 59
Texas Instruments MSP430 microcontroller, 11
Texas Instruments MSP430F1122, 26, 111
Thermal relief, 34, 35, 40, 43, 46, 181
Three Laws of Robotics, 3
Timer, 21, 26, 113, 115, 116, 119, 122, 123, 136, 161, 178, 181
Traces, 34, 35, 43, 45, 46, 47, 51, 53, 60, 105, 106
Transistor, 21, 23, 46, 115, 123, 134, 161, 164, 182
Tripod, 10, 64, 97, 112
Two degrees of freedom, 7, 8, 11, 75, 123, 125, 132, 148, 149, 151, 154, 155, 156
Two-degrees-of-freedom jumper, 103
Two-degrees-of-freedom mode, 103, 107, 115

ULN2803A, 26, 133, 134, 135, 139, 140, 141, 182

Via, 34, 35, 45, 46
Volatile, 182
Voltmeter, 141, 182

Walking robot, 5, 6, 63, 112
Watchdog timer (WDT), 119, 123, 182
Wheeled vehicle robot, 4

About the Authors

James M. Conrad received his bachelor's degree in computer science from the University of Illinois, Urbana, and his master's and doctorate degrees in computer engineering from North Carolina State University. He is currently an associate professor at the University of North Carolina at Charlotte. He has served as an assistant professor at the University of Arkansas and as an instructor at North Carolina State University. He has also worked at IBM in Research Triangle Park, North Carolina, and Houston, Texas; at Ericsson/Sony Ericsson in Research Triangle Park, North Carolina; and at BPM Technology in Greenville, South Carolina. Dr. Conrad is a Senior Member of the IEEE. He is also a member of Eta Kappa Nu, the Project Management Institute, and the IEEE Computer Society. He is the author of numerous book chapters, journal articles, and conference papers in the areas of robotics, parallel processing, artificial intelligence, and engineering education. Dr. Conrad can be reached at the following address:

UNC at Charlotte
Department of Electrical and Computer Engineering

ISBN 0-471-48882-8

9201 University City Blvd.
Charlotte, NC 28223
Phone: 704-687-2535
Fax: 704-687-2352
E-mail: jmconrad@uncc.edu, jconrad@stiquito.com

Pamela Callaway received a bachelor's in both Psychology and Russian from UNC Chapel Hill in 1997, and became interested in computers when she took a programming course for the Psychology degree. She later went on to get a programming certificate from NC State University, and now works at IBM as a Software Developer, as well as pursuing her Master's degree in Computer Science. Her areas of interest include robotics, artificial intelligence, human–computer interaction, and neuroengineering. She can be reached at pcallaway@pobox.com.

James G. Martin taught himself to draw at a very early age. His love for cartooning prompted him to seek formal training and a Bachelor's degree in Illustration at Northern Illinois University. He has designed holiday giftware, art for educational toys, and has worked with many licensed properties such as Disney, The Muppets, and Looney Tunes. He is currently an art director for a leading agency that designs and produces children's meal toys for fast food resturants. James lives happily in Lombard, Illinois, with his wife and son. Mr. Martin can be reached by e-mail at jmartin@stiquito.com.

Andy McClain graduated from UNC Charlotte in 2001 with a Bachelor's in Computer Science. He returned to UNCC from 2003–2004 where he completed work on a Bachelor's in Electrical Engineering. It was during this work for his EE degree when he began work with Dr. Conrad through his class and outside projects. Andy's primary contribution is the C code for the TI-based microcontrolled Stiquito. Andy can be reached at amcclain@stiquito.com.

Jonathan W. Mills received his doctorate in 1988 from Arizona State University. He is currently an associate professor in the Computer Science Department at Indiana University and director of Indiana University's Analog VLSI and Robotics Laboratory, which he founded in 1992. Dr. Mills invented Stiquito in 1992 as a simple and inexpensive walking robot to use in multirobot colonies, and with which to study analog VLSI implementations of biological systems. Dr. Mills is currently researching biological computation in the brain using tissue-level models of neural structures implemented with analog VLSI field computers. Dr. Mills can be reached at the following address:

Indiana University
215 Lindley Hall, Computer Science Department
Bloomington, IN 47405
Phone 812-855-6486; Fax 812-855-4829
E-mail: stiquito@cs.indiana.edu

Rajan Rai completed his bachelor's from REC Bhopal, India, in Electronics and Communication. During his undergraduate years, he did various projects in the computer engineering field like wireless communication between two PCs, CNC machine design using stepper motors, and controlling various remote devices from the desktop. He worked in the IT industry after completing his bachelor's. Currently, he is working toward his Master's Degree at the University of North Carolina at Charlotte in Electrical and Computer Engineering. His fields of interest include embedded systems and communication.

Steve D. Tucker received his bachelor's and master's degrees in electrical engineering from the University of North Carolina at Charlotte. He is currently working on his doctorate degree at the same university. While working on his master's degree, he worked for Litmas, Inc., a start-up company specilizing in radio-frequency power supplies for semiconductor manufacturing. Mr. Tucker is a Student Member of the IEEE and the IEEE Solid-State Circuits Society. His research interests include analog and mixed-signal integrated circuit design, rail-to-rail CMOS amplifiers, low-power and low-noise CMOS design, simulation tools, and analog fault modeling.

Scott Khoi Vu entered college at an early age. At the age fourteen, he was accepted as a part time student at North Carolina State University, and at sixteen, he enrolled as a full time student. Scott is currently working towards his Doctoral of Philosophy in the joint Department of Biomedical Engineering at North Carolina State University and the University of North Carolina at Chapel Hill. He received his Bachelor of Science degrees in Electrical Engineering, Computer Science, and Computer Engineering on May 2004. His interests in robotics led him to meet Dr. James M. Conrad, who helped him with the development of the two degrees of freedom Stiquito. Scott is currently involved with bioinformatics and computational biology research at North Carolina State University and the University of North Carolina at Chapel Hill. Scott can be reached at the follow email address: skvu@ncsu.edu